Zoulikha Zoulikha
Djamila Halliche
Prashant Chaudhari

Methane Reforming with CO_2 in Presence of Catalysts Type HDL

Zoulikha Zoulikha
Djamila Halliche
Prashant Chaudhari

Methane Reforming with CO_2 in Presence of Catalysts Type HDL

ScienciaScripts

Imprint
Any brand names and product names mentioned in this book are subject to trademark, brand or patent protection and are trademarks or registered trademarks of their respective holders. The use of brand names, product names, common names, trade names, product descriptions etc. even without a particular marking in this work is in no way to be construed to mean that such names may be regarded as unrestricted in respect of trademark and brand protection legislation and could thus be used by anyone.

Cover image: www.ingimage.com

This book is a translation from the original published under ISBN 978-620-2-53336-2.

Publisher:
Sciencia Scripts
is a trademark of
International Book Market Service Ltd., member of OmniScriptum Publishing Group
17 Meldrum Street, Beau Bassin 71504, Mauritius
Printed at: see last page
ISBN: 978-620-2-50461-4

Methane Reforming with CO2 in Presence of Catalysts Type HDL

Dédicaces

I dedicate this modest work:

To my God Allah, who has given me the strength to live to this day, the courage, the

health, the will and especially the patience to carry out this

memory EL- HAMDOU-LI-ALLAH.

To the messenger of ALLAH Mohamed QSSL and the Muslim nation.

To those who I love very much in my life:

My dearest Mother, who sacrificed her life to raise me, who gave me everything.

She's the reason I became the person I am today.

My dearest Father, whom I love very much, whom I respect very much ...

To my sisters, especially Brother Zohra.

To my brother Mohamed.

To my country, Algeria.

To all those who are very, very dear to me...

A. Zoulikha

ACKNOWLEDGEMENTS

The research work presented in this thesis was carried out at the Natural Gas Chemistry Laboratory of the Faculty of Chemistry of the University of Science and Technology Houari Boumediene (U.S.T.H.B) under the direction of Ms. **D. Halliche, Senior Lecturer, U.S.T.H.B.**

I thank my **almighty God ALLAH** *who gave me the courage and the will to come to the end of this work.*

First of all, I would like to warmly thank Mrs C. RABIA, Professor (U.S.T.H.B) for having welcomed me in her laboratory and for having been available to offer many advices. I thank her all the more for her availability and her many pieces of advice.

I would also like to thank Ms. D. HALLICHE Maître de Conférences (U.S.T.H.B) for the quality of her supervision, her wise advice, her encouragement and the many discussions which enabled me to carry out this work.

I would like to warmly thank Mr. A. BALIOUAMER Professor at the Faculty of Chemistry (U.S.T.H.B) for having done me the honour of chairing the dissertation jury.

I was very honoured by the participation of Professor M. TRARAI (U.S.T.H.B) in my dissertation jury.

I would also like to thank Mr. D. NIBOU, Senior Lecturer (U.S.T.H.B) for having accepted to be part of my thesis jury.

I would like to thank Mr. K. BACHARI Research Master (C.R.A.P.C- U.S.T.H.B) for his help and his many advices. I also thank him for having accepted to be part of this jury.

I also thank Mrs O. CHERIFI, Professor (U.S.T.H.B), for her advice and encouragement throughout my work.

I would like to thank Mr F. very warmly. CHERIFI in charge of the chemical analysis service of the CETIM-BOUMERDES centre for his welcome in his laboratory to carry out the different analyses, I also thank his team in particular N. ZABOUT and H. TALAHARI.

I would like to thank Mr. A. warmly. SAADI for his help.

I warmly thank all my laboratory colleagues for their sympathy and presence: MERABTI Razika.

I would also like to thank all the people in the Faculty of Chemistry who helped me to carry out this work.

Introduction
General

GENERAL INTRODUCTION

Today there is no longer any room for doubt, the atmosphere is getting warmer... The fever is rising, as in a living organism, a sign of serious illness (figure-1)! Throughout its history, the Earth has experienced periods during which greenhouse gases, of volcanic origin, have produced situations rich in lessons about what current or future climatic disturbances can hold for us.

Figure-1: Fever in the Earth's organism.

Emissions of greenhouse gases, including carbon dioxide, increased by 20% between 1990 and 2004 according to a new index published by the federal oceanographic and atmospheric agency.

The indicators measured in the 20th century show clear trends:

- The temperature increased by 0.6°C ± 0.2°C, which is considerable.

- Sea levels have risen 10 to 20 cm.

- Northern Hemisphere polar ice has decreased by 10%.

Monitoring the climate over the next 20 years (Table-1) then becomes vitally important, necessitating the implementation of new climate-dependent insurance strategies.

Table-1: Evolution of atmospheric CO_2 concentration and climate change.

Period of occurrence of maximum CO_2 emissions	Atmospheric CO_2 concentration	Global temperature increase relative to the Temperature of 1850
2000 à 2015	350 to 400 ppm	2.0 to 2.4 °C
2000 à 2020	400 to 440 ppm	2.4 to 2.8 °C
2010 à 2030	440 to 485 ppm	2.8 to 3.2 °C
2020 à 2060	485 to 570 ppm	3.2 to 4.0 °C
2050 à 2080	570 to 660 ppm	4.0 to 4.9 °C
2060 à 2090	660 to 790 ppm	4.9 to 6.1 °C

In this context, natural gas remains the safest and cleanest source of energy among fossil fuels, hence the need to involve it more in the various economic and industrial sectors.

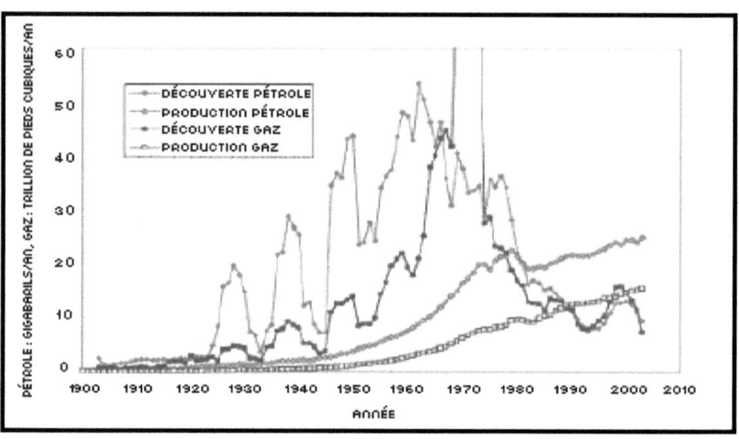

Figure-2: Annual World Oil and Gas Production.

Moreover, as scientific and technological progress is constantly evolving, it has become the prerogative of the great nations whose objectives are to maintain economic independence and master new technologies that could make it possible to move away from the binary choice between clean air and the development of fossil fuels.

Indeed, natural gas, with an abundance quantified by world production (Figure-2) and a very high storage capacity, offers great potential for exploitation as an important energy source [1-3].

The most developed industrial processes today use as a first step the formation of synthesis gas (CO $/_{H2}$). We can quote :

- ➢ Methane vapor reforming : $\qquad CH4 +_{H2O} CO \longrightarrow + 3_{H2}$ (1).
- ➢ Partial oxidation of methane: $CH4 + ½ O CO \longrightarrow + 2_{H2}$ (2).
- ➢ Dry methane reforming : $\qquad CH4 +_{CO2} \longrightarrow 2\ CO + 2_{H2}$ (3).

Syngas is also one of the new and renewable energies and is considered the fuel of the future in terms of the development of fuel cells, power plants and new non-polluting vehicles.

One suitable means of producing synthesis gas is dry methane reforming (Eq.3). From an environmental point of view, this reaction can contribute to the decrease in atmospheric concentrations of CO2 and CH4, considered as greenhouse gases par excellence. In addition, this reaction offers the opportunity to have an H2/CO ratio close to unity, a value that is much sought-after for the manufacture of ammonia and other industrial applications. The dry methane reforming reaction requires relatively high temperatures. Thus, the main objective of this work is to develop catalysts that are active at the lowest possible temperatures to avoid coking of solids at high temperatures.

A wide variety of catalysts have been tested for this reaction. In addition, as the dispersion of the active phase in the catalyst is an important parameter in the catalytic act, hydrotalcite-type materials are used as catalysts for the dry methane reforming reaction.

Hydrotalcites are mixed lamellar double hydroxides of divalent and trivalent metals. Their structure of general form $[M1\text{-}x^{2+} Mx^{3+} (OH)_2]^{x+} [^{An\text{-}}]_{x/n} . m_{H2O}$ is based on a stack of sheets of composition $M(OH)_2$ similar to those of brucite $Mg(OH)_2$ [4].

The large specific surface area of these materials, as well as good dispersion of the active phase, together with their anionic exchange property and basic character, justify the growing interest in this type of material.

Hydrotalcites before and after heat treatment are used in many catalytic applications. They also find applications in the medical field or as an adsorbent in chromatography.

This manuscript is divided into four chapters:

> The first chapter is devoted to the bibliographical study where it will be question of recalls concerning materials of hydrotalcite type in a first place and, in a second place it will be question of the reaction of dry reforming of methane.

> The second chapter lists and describes the different physico-chemical analysis techniques used during our study.

> In the third chapter, we will present the hydrotalcite-type catalysts and their method of development as well as their main characteristics. We will also try to determine the nature and position of the cationic and/or metallic nickel species responsible for the activity.

> In the fourth chapter, we will undertake a systematic study of catalysts and the optimal conditions for their operation will be deduced in a first step.

In a second step, the Friedel-Craft reaction will be discussed, for which we will test catalysts that have shown total catalytic inertia for the dry methane reforming reaction. We will conclude with a general conclusion.

BIOGRAPHICAL REFERENCES

Rostrup-Nielsen, Catal. Today, 63 (2000) 159.

J.R. Rostrup-Nielsen, Top. Catal, 1 (1994) 377.

3]-J. H. Lunsford, Catal. Today, 63 (2000) 165.

F. Cavani, F. Trifirò, A. Vaccari, Catal. Today, 11 (1991) 17.

Chapter I
Bibliograph ical study

CHAPTER I

FIRST PARTY

HYDROTALCITES, CATALYTIC MATERIALS.

INTRODUCTION

Thanks to their anionic exchange property, their basic character, their strong adsorbent character etc ..., lamellar double hydroxides have been the subject of growing interest since the end of the sixties. These materials find applications in various fields, particularly in the field of heterogeneous, pharmaceutical and even environmental catalysis.
 Great efforts were then made to better characterize this type of material and to improve the methods for its development.

I. HISTORICAL CONTEXT

The history of hydrotalcite began in 1842 in Sweden where these solids were identified as a naturally occurring mineral consisting of aluminum and magnesium carbonate [1]. Moreover, Mannasse was the first research mineralist to present the general formula of hydrotalcite: **Mg6 Al2 (OH) $_{16}$. CO3. 4 $_{H2O}$.**

He was also the first researcher to recognize the importance of carbonate ions in the construction of the hydrotalcite framework. In 1942, Feitcknecht synthesized several families of hydrotalcites and called them "doppelschichtstrukturen" (double layer structure), which means "double lamellar structure". Further studies led him to affirm that the structure of these materials consists of layers of hydroxides of a single cation, intercalated by a second layer containing the second cation, (Figure-1) [1,2]. This structure was categorically rejected by Allman and Taylor, but they finally recognized, much later, that the structure of these solids is based on a stack of sheets that resembles brucite of $Mg(OH)_2$ compositions [1,2].

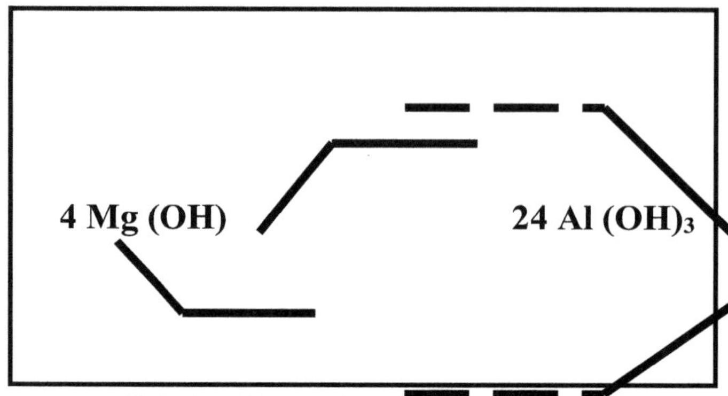

Figure-1: Structure of hydrotalcite, according to Feitcknecht (1942) [1].

The introduction of hydrotalcites in the field of catalysis will take place for the first time thanks to the work of Zelinski and Kommarewsky [1].

Years later, Molstad and Dodge were able to develop zinc-chromium hydrotalcites that they tested for methanol synthesis [1].

Later, in 1977, Miyata et al. published a paper on anionic clays of the hydrotalcite type and described them as naturally abundant and easily synthesized in the laboratory [1].

II. GENERAL INFORMATION ON HYDROTALCITES

1. Introduction

Hydrotalcites are anionic clays with a double lamellar structure containing exchangeable anions. The structure of the double lamellar hydroxides is based on a stack of sheets of $M(OH)_2$ composition similar to those of the brucite $Mg(OH)_2$ shown in Figure-2 [1,3]. Brucite $Mg(OH)_2$ consists of octahedra with Mg^{2+} ions at the vertices.

The general chemical formula of hydrotalcites (1) is:

M1-x2+ Mx3+ (OH) $_2$]$^{x+}$ [$^{An-}$] $_{x/n}$ .m $_{H2O}$ (1)

Where:

M2+ , M3+ : Divalent and trivalent cations located in the octahedral locations in the hydroxylated layers.

x : The ratio (n (M3+)/ n (M2++M3+)).
^{An} : is an exchangeable anion of the
 interlamellar layer.

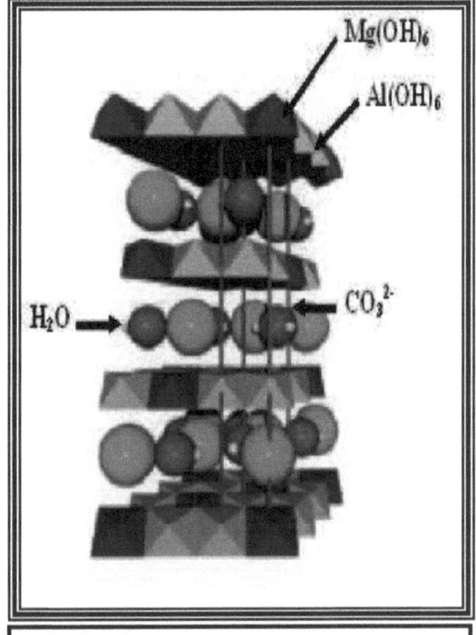

Figure-2: Structure of hydrotalcite

Overall electrical neutrality is provided by anionic species interspersed in the interleaf space, together with water molecules 1,4]. Anions and water molecules are located in a disordered way in the interleaf space [1,2].

The structure of hydrotalcite crystallizes in two systems of crystallographic symmetry, hexagonal and rhombohedral [1,2].
The structural parameters and properties of hydrotalcites are strongly influenced by the nature of the interleaf anions and the value of the ratio [n (M3+)/ n (M2++M3+)] [5].

2. General formula of hydrotalcite and its chemical composition
The above formula (1) shows the chemical atomic content of the two hydrotalcite polytypes (Hexagonal and Rhombohedral). Some work has shown that pure natural hydrotalcite has an x value of 0.25 with carbonates as interlamellar anions [1].
The nature of M (II), M (III) and the values of x and m will be discussed later.

a. Nature of the sheets

Many divalent and trivalent cations can be combined to form lamellar double hydroxides. The most commonly synthesized flakes are based on magnesium and aluminum, as in natural hydrotalcite. However, other cations may be combined.

-Cations divalent: Zn^{2+}, Ni^{2+}, Cu^{2+}, Ca^{2+}, Mg^{2+}, Mn^{2+}, Cr^{2+}...

-Trivalentations: Cr^{3+}, Fe^{3+}, Mn^{3+}, V^{3+}, Ga^{3+}, Al^{3+}, Ni^{3+}...

However, the obtaining of the hydrotalcite structure is limited by the ionic rays of the M^{2+} and M^{3+} cations, so that these must be close to the value $0.65A°$ (characteristic of the Mg^{2+} ion in natural hydrotalcite) [1.2]. In Table-1 we report the ionic radii of some bivalent and trivalent cations that have the possibility to occupy octahedral sites and lead to the double lamellar structure. The literature reports that all cations since then, Mg^{2+} to Mn^{2+} can lead to a hydrotalcite-like structure [1].

It should nevertheless be noted that Be^{2+} ions, because of their small size, cannot occupy the octahedral sites of brucite and that Ca^{2+} ions have a sufficiently high ion radius not to give a stable structure to hydrotalcite. In fact, these cations will probably form a structure other than that of brucite [1].

Furthermore, all trivalent ions form the structure of hydrotalcite with an ion radius between 0.5 and 0.8 $A°$ with the exception of the cations V^{3+} and Ti^{3+}, of which these ions are not stable [1,2].

Table-1: Values of ionic radii ($A°$) to obtain the hydrotalcite structure [1].

M (II)	Be	Mg	Cu	Nor	Co	Zn	Fe	Mn	Cd	Ca
	0,30	0,65	0,69	0,72	0,74	0,74	0,76	0,80	0,97	0,98
M (III)	Al	Ga	Nor	Co	Fe	Mn	Cr	V	Ti	In
	0,50	0,62	0,62	0,63	0,64	0,66	0,69	0,74	0,76	0,81

b. The values of x

The structure of hydrotalcite can exist for values of x in the range (0.1-0.5). However, some work has shown that it is possible to obtain it only in the range $0.2 \leq x \leq 0.33$ [1, 6.7].

c. The values of m

m denotes the number of water molecules located in the interleaf space. The quantity of water is thus determined by thermogravimetric analysis by measuring the weight loss [1.8].

d. Nature of the anions in the interleaf space

In general, it is difficult to have an accurate structural description of the interleaf domain. It can also be said that the interlaminar space is a highly disordered environment [1,2]. Nevertheless, in the case of simple entities such as carbonates or chlorides, the anions statistically occupy well defined sites.

The interlayer space and the physico-chemical properties of the material are defined by the nature of the anions that constitute it. At present, no limits concerning the nature of the interleaved anions have been demonstrated. However it is necessary that :

- These anions are stable under the operating conditions.
- That they have no steric or geometric constraints.

A wide variety of anions can be inserted in the intersheet space:

- Simple anions: $NO3-$, $OH-$, $CO32-$, $[Fe(CN)_6]^{4-}$ ...
- Heteropolyacids (heteropolyoxometalates): $(PW12O40)^{3-}$...
- Oxometallates : Chromates, vanadates, molybdates...

The presence of interleaf anions leads to polytypes of hexagonal (2H) and rhombohedral (3R) symmetry. It is clear that the molecular structure of the anion and its orientation in the interlamellar space influence the type of stacking.

3. Mineral polytypes of hydrotalcites

The great variety of trivalent and divalent cations justifies the existence of mineral polytypes of hydrotalcites such as: Takovite, Stichtitte, Pyroaurite (Table-2, Figure-3).
1,2 .

Table-2: Examples of natural polytypes of hydrotalcite [1].

No and	chemical composition	symmetry
Hydrotalcite Mg6 Al2 (OH)$_{16}$ CO3. 4 $_{H2O}$		3R
Manasseite Mg6 Al2 (OH)$_{16}$ CO3. 4 H2O		2H
Pyroaurite Mg6 Fe2 (OH)$_{16}$ CO3. 4.5 $_{H2O}$		3R
Sjogrenite Mg6 Fe2 (OH)$_{16}$ CO3. 4.5 $_{H2O}$		2H
Takovite Ni6 Al2 (OH)$_{16}$ CO3, OH. 4 $_{H2O}$		3R
Reevesite Ni6Al2 (OH)$_{16}$ CO3. 4 $_{H2O}$		2H

Figure-3: Mineral polytypes of hydrotalcite, (A) Natural pyroaurite.
(B) Natural Takovite.

III. METHODS OF PREPARATION

1. Introduction

Anionic clay-type hydrotalcites are less diffused in nature than cationic clays [1,9,10]. For this reason, much work has been done in the field of hydrotalcite synthesis. The ease of preparation and the relatively easy experimental conditions encourage research work in this field [1,9,10].

1,2 .

The construction of the hydroxyl framework is a process based on the controlled precipitation of aqueous solutions or suspensions, containing both the metal cations intended to take place in the hydroxyl framework and the anions intended to occupy the interlamellar domains.

The nature and texture of the phases obtained depend strongly on the operating conditions, including temperature, rate of addition of reagents, concentration of solutions, pH of solutions, nature of intercalated anions, molar ratio of divalent and trivalent cations ($M2+/M3+$) [2]. As shown in Table-3.

There are two essential preparation methods: the classical co-precipitation method [1,2,10], and anion exchange.

Table-3: Factors influencing hydrotalcite synthesis.

Structural variables	Preparation variable
Cation size	pH
Value of x	Concentration of cations and anions in solution
Cation stereochemistry	Temperature and stirring speed
Cationic mixture	Presence of impurities
Nature of the anion	---

2. Co-precipitation

Co-precipitation (solid phase crystallization spc method) is the most widely used method for preparing hydrotalcites [9, 11].

Co-precipitation is the contamination of a precipitate with substances normally soluble under the conditions of precipitation. It consists in causing the simultaneous precipitation of divalent and trivalent metal cations by the addition of a basic species in adequate proportions. A slow addition of the reagents is generally favourable to a good organisation of the prepared phase as reported by Miyata et al [1].

3. Anion exchange

The anion exchange reaction is a feasible reaction for hydrotalcites, i.e. the ion-covalent structure of the sheets is retained, while weaker anion-sheet bonds are destroyed.

The anion exchange operation consists of contacting the precursor of the lamellar double hydroxides generally containing carbonate, chloride or nitrate ions with a solution containing the anion to be intercalated [1,12].

4. Rebuilding of the framework " MEMORY EFFECT

After a careful calcination of the hydrotalcite, it is always possible to restore its framework by simple rehydration in the presence of a selected anionic species. This is known as the **memory effect of** hydrotalcites. This would be a true reconstruction of the hydrotalcite structure [1,2,13].

IV. PROPERTIES OF HYDROTALCITES

1. Obtaining mixed oxides

The calcination of hydrotalcites leads to their dehydration and then to dehydroxylation and decarboxylation which is accompanied by the collapse of the lamellar structure. Calcination can lead to mixed oxides when the temperature is sufficiently high. The experimental conditions of this heat treatment confer certain properties to the materials obtained [1, 2, 10] including :

-Large specific surface area (100-300 m2/g) (after dehydration, dehydroxylation, decarboxylation).
-Basic, acidic and even redox properties.
-Good dispersion of the active phase.
-Memory effect.

2. Basicity of hydrotalcites

The basic properties were demonstrated by adsorption of acidic probe molecules such as CO_2, SO_2 [14].

The basic character has been examined in the case of the compound MgAlCO3-HT, whose heat treatment makes it possible to obtain oxides of the MgO type. The basic properties of MgO have been widely studied and shown in the literature [1].

Strong basic sites are assigned to O2- species, while medium basic sites are assigned to (O-) groups located near hydroxyl groups. Another type of weak basicity is assigned to (OH-) groups (Bronsted basicity) [1].

4. Memory effect

As already mentioned, hydrotalcite leads after calcination to a set of mixed oxides. The structure of the basic hydrotalcite can, however, be restored after appropriate treatment (see Figure-4), if the calcination took place at a relatively low temperature (around 450°C) [1, 2, 15].

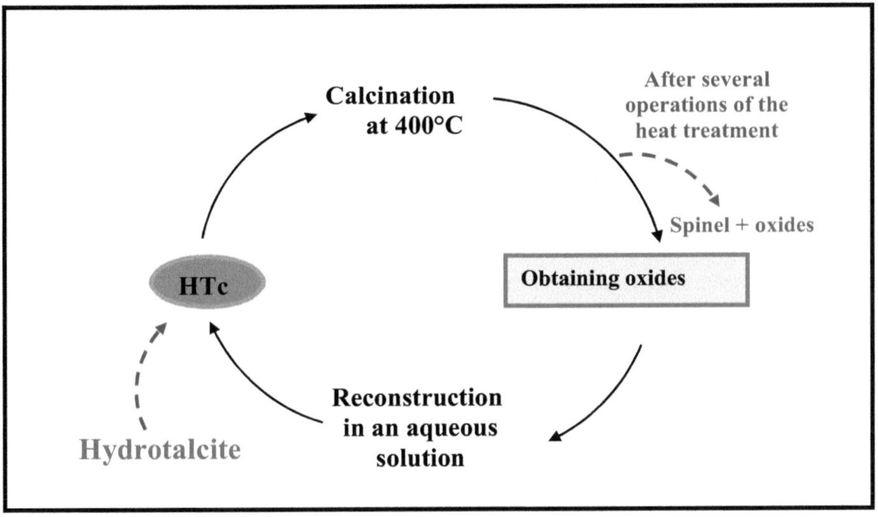

Figure-4: Thermal cycle of the different stages of the reconstruction of the hydrotalcites.

When calcined hydrotalcites are rehydrated in a nitrogen-saturated solution, the structure of these solids rebuilds. We call this property the "memory effect of calcined hydrotalcites" [1].

V. HYDROTALCITE BENEFICIATION

1. Introduction

The numerous textural and structural properties of hydrotalcites (calcined or not) justify their numerous uses in hitherto unsuspected fields (medicine, catalysis, industry ...) [1, 2].

Hydrotalcites are often used as catalytic precursors thanks to [15] :

-The ability to vary divalent and trivalent cations.

-The good dispersion of these cations throughout the structure of the material

-The obtaining of thermostable materials with relatively high specific surfaces.

The applications of hydrotalcites are numerous both in the field of heterogeneous and homogeneous catalysis.

These materials are thus used as catalysts for several types of chemical reactions such as, the production of hydrogen by vapor-deformation of vegetable oils (Soya, Corn, Sunflower) [16], conversion of isopropanol [17] ... etc.. They also find applications in the medical field or as adsorbents in the field of chromatography [1].

As we can see, the prospects opened up by the use of hydrotalcites go beyond catalysis and adsorption alone. New themes are emerging, such as the separation of biologically active molecules (asymmetric synthesis), the development of materials for the electrochemical industry (electrode), the design of sensors, etc... These new uses require the collaboration of researchers from very different disciplines, leaving chemists with the primary mission of producing materials whose textural characteristics and surface properties are perfectly mastered. Figure-5 shows the different applications of our solids.

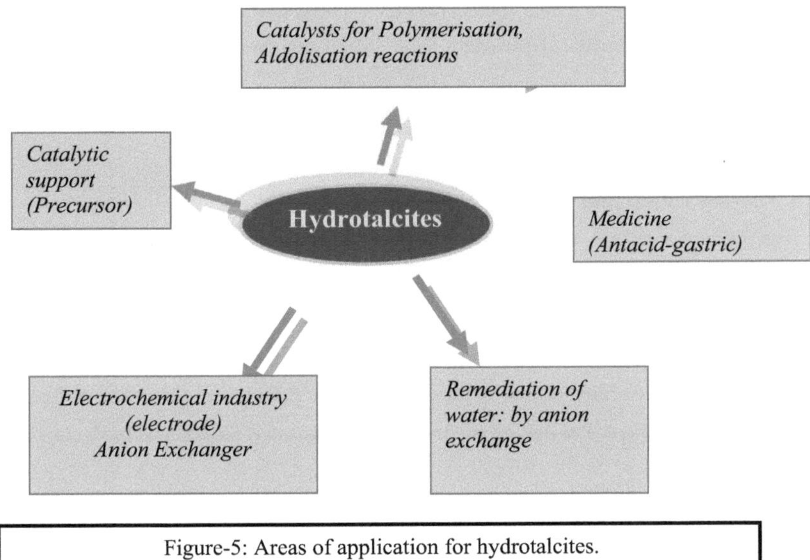

Figure-5: Areas of application for hydrotalcites.

2. Catalysis field

Hydrotalcites have several catalytic applications, which have been reported in the literature.

The basic mixed oxides obtained after heat treatment (at 400°C) of hydrotalcites showed interesting catalytic properties in heterogeneous catalysis. As is the case in polymerization, dehydrogenation, alkylation and aldolization reactions [1,18]. These materials are also applied to methane reforming reactions [1, 19, 20].

3. Environmental field

The involvement of double lamellar hydroxides in the environmental field can be considered as follows:

-hydrotalcites can be used to trap organic or inorganic chemical pollutants.

-hydrotalcites can also be used as a carrier or catalyst for the degradation of chemical pollutants.

On the other hand, with a high anion exchange capacity and a highly reactive hydroxylated surface, double lamellar hydroxides are interesting materials for the trapping of ecologically undesirable molecules. A study has also been carried out on the possibility of intercalation of polluting molecules such as chromates by this type of material.

4. Pharmaceutical field

Hydrotalcites also have applications in the pharmaceutical field 1. 1. Currently, the pharmaceutical chemical formula is used on a hydrated hydrotalcite of type MgAl-HT basic, whose anions are carbonates. Its pharmacological property is based on its antacid character which opposes the lowering of the intra-gastric pH. However, this material presents certain contraindications in the case of severe renal insufficiency and children under 15 years of age. The dosage and method of administration declared by doctors is the usual oral dose.

5. Modified electrode design

Hydrotalcites possess in addition to a high anion exchange capacity, a homogeneous and regular dispersion of the positive charges of the sheets and the negative charges of the interleaf spaces 21. For all these reasons, these materials are offered for possible use in electrochemistry as electrodes. The application principle consists in immobilizing on an electrode electroactive molecules in order to improve and accelerate electronic transfers in electrostatic chemical reactions. Some authors then proposed a model (figure-6) of electrostatic energy (**EPEM**) 21] which is based on :

- The attraction between the positive charges of the layers and the intercalated anions.
- The repulsion between the anions in the inter-lamellar space.

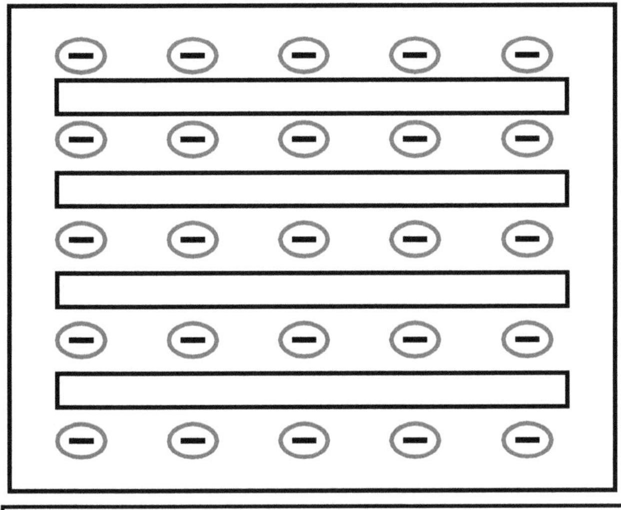

Figure-6:   Model of the   electrode based on the structure of the

hydrotalcite (EPEM).

PART TWO: CATALYTIC APPLICATIONS
A/ DRY METHANE REFORMING REACTION

INTRODUCTION

Natural gas is a natural hydrocarbon. Hydrocarbons are a class of organic compounds consisting of carbon and hydrogen. They include crude oil, natural gas and coal.

Natural gas is a low-polluting energy, whose environmental qualities are recognized and appreciated. In the face of growing environmental constraints, natural gas can only be favoured over other, more polluting fuels. Its use in power plants or in combined heat and power systems offers better efficiency than other energy sources and allows a reduction in investment and operating costs. Extensive studies on the moderate consumption of natural gas show that world consumption of natural gas reached 2120.10^9 m^3 in 1991 and about 2565 to 2725.10^9 m^3 in the year 2000. In 2020, it could reach 3100 to 3500.10^9 m^3.

The main component of natural gas is methane. It is a very stable molecule with low reactivity. It can also contain light hydrocarbons (C_2, C_3, C_4) and H_2S (hydrogen sulphide). Unlike distillation residues from crude oil, it is easily purified by desulphurisation on catalytic beds based on cobalt, zinc and molybdenum. Natural gas, on the other hand, contains no heavy metals or aromatic or naphthenic compounds.

I. KEY VALUE DRIVERS

The conversion of natural gas into synthesis gas (CO/H2) offers varied and economically interesting opportunities. This indirect way of valorising natural gas is currently attracting great interest from the oil industry, as shown by *SHELL*'s investments, for example. Among the direct natural gas upgrading routes (Figure-7), the conversion of methane by oxidation into methanol is one that has been widely explored.

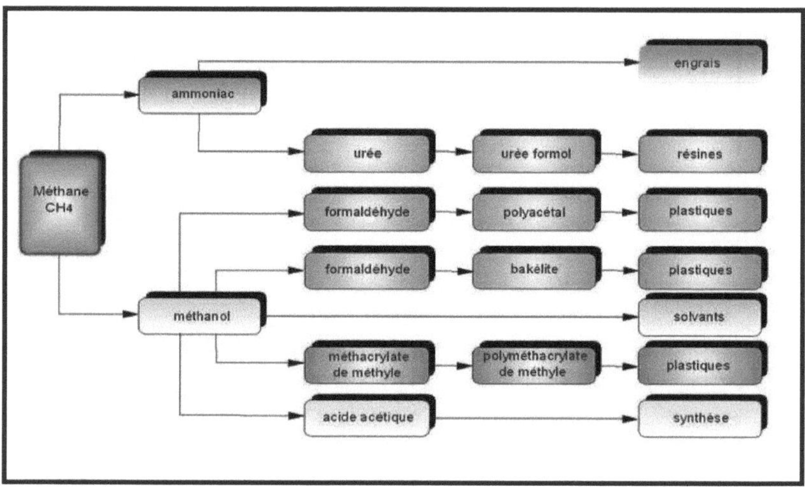

Figure-7: Methane Upgrade Pathways.

However, process performance remains poor: if conversion is high, then selectivity is low and vice versa. In addition, methanol is more reactive than methane with respect to oxygen. Another direct route for processing natural gas has been developed and worked on since 1982. This is the oxidative coupling of methane. In this case, conversion rates can reach 25% with ethane-ethylene selectivities of up to 90%. Nevertheless, these performances are still insufficient for the risk of a first industrial implementation to be taken. Most current direct conversion technologies are faced with the problem of the chemical inertia of methane and the unfavorable thermodynamics of its transformation. The synthesis gas route seems to be the only viable route in the short to medium term.

Algeria has a large reserve of natural gas used mainly as a source of industrial and domestic energy. Algerian gas contains more than 80% methane, the rest being heavier hydrocarbons.

The chemical gas industry consists mainly of the transformation of gas into ammonia and methanol via synthesis gas. Indeed, synthesis gas (H_2, CO) is produced by exploiting methane, according to the following catalytic processes:

- Methane vaporeforming: $CH_4 + H_2O \longrightarrow CO + 3\,H_2$ (1).
- Partial oxidation of methane: $CH_4 + \tfrac{1}{2}\,O_2 \longrightarrow CO + 2\,H_2$ (2).
- Dry methane reforming : $CH_4 + CO_2 \longrightarrow 2\,CO + 2\,H_2$ (3).

The most recent data in the literature indicates that syngas compounds could become important feedstocks in the long term. The interest in these compounds is all the greater because they can be obtained from any source of carbon and hydrogen. Hydrogen is a strategic chemical agent used in the nitrogen fertilizer industry and in petrochemicals in hydrogenation and cracking reactions.

On the other hand, in recent years, much attention has been paid to the reaction (3) for the environmental impact it has (it leads to a decrease in the concentration of two greenhouse gases in the atmosphere (CH_4, CO_2). It is thus used together with carbon monoxide methanation in the storage and transport of solar energy.

II. SYNTHESIS GAS

The synthesis gas is obtained by the following important reactions:

> **Catalytic steam reforming**

$$CH_4 + H_2O \longrightarrow CO + 3 H_2 \quad (1). \qquad \Delta H°298 \quad = \quad +206$$
$$KJ/mole. \quad CO + H_2O \longrightarrow CO_2 + H_2. \quad (1'). \qquad \Delta H°298 = -41 \text{ KJ/ mole.}$$

This reaction is always accompanied by the gas-to-water conversion reaction.
(1'). This reaction takes place in the presence of nickel-based catalysts.

> **Partial oxidation**

$$CH_4 + \tfrac{1}{2} O_2 \longrightarrow CO + 2 H_2 \quad (2). \qquad \Delta H°298 = -39 \text{ KJ/ mole.}$$

This reaction can be carried out at high temperature (1000-1200°C) in the absence of a catalyst. However, significant research is currently being carried out on the development of catalysts that allow the reaction to be carried out at lower temperatures.

> **Dry methane reforming**

$$CH_4 + CO_2 \longrightarrow 2 CO + 2 H_2 \quad (3). \qquad \Delta H°298 = +246 \text{ KJ/ mole.}$$

This reaction is generally followed by the reverse reaction of gas with water

$$(R.W.G.S). \quad CO_2 + H_2 \longrightarrow CO + H_2O \quad (4). \quad \Delta H°298 = -41 \text{ KJ/ mole.}$$

The dry methane reforming reaction is a reaction with a high environmental impact, as it minimizes the concentration of two greenhouse gases. The process of dry methane reforming to produce synthesis gas ($CH_4 + CO_2 \longrightarrow 2CO + 2H_2$), is of particular interest given the role that this response plays in reducing air pollution through the use of two dreaded greenhouse gases. This process is industrially interesting since it allows to obtain an H2/CO ratio close to unity, an ideal value for Fischer-tropsh synthesis [22].

Furthermore, the dry methane reforming reaction is a reaction whose selectivity and producibility is highly dependent on the nature of the catalysts used.

III. CATALYSTS FOR DRY METHANE REFORMING

One of the major challenges in the commercialization of the CH_4 / CO_2 process is the development of catalysts that allow the reaction to take place without significant deactivation by carbon deposition. All the elements of group VIII (Rh, Pt, Ni, Co, Fe, Ir...) with the exception of Osmium) are known to be more or less active in the dry methane reforming reaction [23,24]. For this reason, some work is focused primarily on the development of catalysts that are active but also resistant to deactivation by carbon deposition [25].

Among the elemental catalysts mentioned above, Rh is by far the one with the best catalytic performance without showing a strong tendency to deactivate by coke deposition [26-28].

However, economic constraints have stimulated and amplified research on low-cost systems such as Ni, Co and Fe supported systems, which represent a fairly good alternative to rhodium for this reaction [29].

Some nickel-based catalysts may even exhibit activities comparable to those obtained with noble metals, particularly NiO-MgO systems, which lead to high activities with a low deactivation tendency [30,31-33].

Furthermore, it is generally accepted that bimetallic catalysts have higher activity and performance than monometallic systems [34].

Crisafulli et al [35] found that the stability of catalysts supported on SiO_2, evolves in the direction Ru > Ni >> Pd. These results are in perfect agreement with the literature data which state that noble metals are more resistant to cockage than nickel alone, probably because carbon is not dissolved in these metals [26,27]. In revenge, Pd would be the only noble metal that strongly deactivates by carbon formation, probably due to the formation of a solid solution of interstitial carbon [24]. In the same study, Crisafulli showed that the addition of Ru to the Ni/SiO_2 system significantly enhances the catalytic performance both in terms of activity and stability over time, however, the addition of Pd to the same base system only slightly enhances the catalytic activity with a very strong tendency to deactivate, compared to the Ni/SiO_2 catalyst [35].

In another study [36], we examined the effect of a metallic addition (Fe, Co, Cu, Ce) to a Ni/α-$Al2O3$ base catalyst. We found that the performance

Catalysts evolved in the following sequence: Ni > Ni - Co > Ni- Cu > Ni - Fe.

In the same context of diversification of catalyst systems, Bhat [37] tested Rh-based catalysts introduced in HY and NaY zeolites. He noted that the zeolite supports could retain large dispersions of the active phase and provided high catalytic performance without any deactivation even after 30 working hours under reaction mixing.

In addition, we have tested a series of catalysts based on zeolite Y (USY) exchanged by a series of transition metals Co, Cr, Fe and Ni [38]. We found that, with the exception of NiUSY, all the catalysts remained inactive with respect to the reaction mixture.

The literature has also reported that Ni-based catalysts deposited on a pentassil-type zeolite (ZSM-5) offer an excellent alternative for the methane reforming reaction [39]. Moreover, modified with K and Ca, the catalyst has significantly improved catalytic performance, i.e., activity is maintained beyond 140 working hours at 800°C without any sign of deactivation.

In addition, a new concept for catalyst preparation has been developed, '*solid phase crystallization (Spc)*'. Using this technique, researchers have succeeded in developing catalyst systems with a highly dispersed phase [9].

Another new class of catalysts based on modified nickel hexaaluminates $ANiAl11 O19_{-a}$ (A= Ca, Sr, Ba, La) is prepared by decomposition of nitrates and calcination at high temperature, seems to provide better performance and stability [40,41].

Hydrotalcite-based catalysts have also shown good catalytic performance in the presence of the CH4/CO2 reaction mixture [1].

For example, some work on NiMgAl-HTc hydrotalcite-type NiMgAl-HTc catalyst has shown that the catalytic behaviour depends mainly on the M(II)/M(III) ratio, the experimental pre-treatment and development conditions (pH, calcination, reduction) [1.40].

In other work on NiAl-HTc hydrotalcite-type NiAl-HTc catalyst, it has been confirmed that the introduction of Mg into the hydrotalcite structure improves the catalytic stability of the material [42].

In the same context, *Casenave* et *al* [14], studied the acid-base properties of mixed oxides derived from NiMgAl-HTc hydrotalcite, by the method of adsorption microcalorimetry of probe molecules where it was shown that the basic character of the sample increased with the Mg content introduced into the NiAl-HTc system and thus improved its catalytic performance in dry methane reforming reaction.

On the other hand, *Bhattacharyya* et *al* [43], compared a hydrotalcite-type NiMgAl-HTc catalyst and a conventional Ni/Al2O3 catalyst. It has been shown essentially that, a better dispersion of the active phase is observed on hydrotalcite-derived samples and therefore a better stability in time and better results are obtained compared to the classical catalyst. The results of *Hou* and *Yashima* [44] are similar to those obtained by *Bhattacharyya*.

IV. SUPPORT EFFECT

Numerous studies have shown that the choice of a support is a determining factor that can influence the activity of a catalyst but also its stability and resistance to coking. The most commonly used support for dry methane reforming is α-Al2 O3 [45]. Other supports such as MgO, TiO2 , SiO2 and La2O3 are also frequently cited in the literature. Recently, great interest has been given to mixed oxides such as Ce1 $_{-x}$ Zrx O2 , LaBO3 (B =Co, Ni, Fe and Cr) and La2 NiO4 both as catalysts and catalyst supports [46]. In this context, Hayakawa et al [47] have prepared nickel-based Ni/BaTiO3 catalyst masses. They found excellent catalytic performance for the dry methane reforming reaction.

An adequate catalytic support must be able to withstand the application of high temperatures and maintain the dispersion of the metal during the various operations of the catalytic act. Bhattacharyya and Chang have recently proposed the use of nickel aluminate spinel catalysts to reduce carbon deposition [48].

In addition, MgAl2O4 magnesium aluminate spinels offer an excellent combination of properties, including a high melting point (2135°C) and chemical and mechanical stability [49].

The importance of the choice of support was also highlighted by the work of Ferreira-Aparicio et al. who found two different sequences for different supports [50] :

on Al2O3: Rh > Ni > Pt > Ru >Ir > Co.

on SiO2: Ni > Ru >$_{Rh>}$Ir > Co > Pt.

The work of Uchijima et al [51], Nakamura et al [52] goes in the same direction and showed that at 620°C, Rh deposited on Al2O3 was 18 times more active than when it is deposited on SiO2. They led to the following sequence of decreasing activities:

Rh/ Al2O3 > TiO2 >$_{/SiO2.}$

However, Uchijima et al [51] state in the same study that the catalyst supported on SiO2 could be improved by adding Al2O3 or TiO2.

Tsipouriari [54], for his part, relates the catalytic performances rather to the size of the Rh crystallites. He notes the following catalytic evolution:

YSZ > Al2O3 > TiO2 > SiO2 > MgO.

Using Pd deposited on different supports as a catalyst, the catalytic activities follow the following sequence: TiO2 > Al2O3 > SiO2 > MgO.

Zhang et al [54] confirm this fact and demonstrate that very often the catalytic activity is related to the nature of the support as well as to the size of the Rh particles. In other works, Tsipouriari et al. state that catalytic activities are strongly dependent on the support and show that the addition of CaO to alumina gives better stability to the nickel-based catalyst [53]. These findings are consistent with those of Bradford et al. [55].

Many working groups have focused their attention on TiO2 and ZrO2 supports. They have found that on these supports, Pt achieves much higher activities than when supported by Al2O3 and SiO2 [56].

Bradford and Vannice have studied the Pt/TiO2 catalyst more closely [55]. They found good performance, which they attribute to the formation of a metal-support interaction that can promote the dissociation of CH4 and $_{CO2}$.

Lercher et al [57,58] studied a Pt catalyst deposited on TiO2 as well as on Al2O3 and ZrO2. They found that among these three supports, ZrO2 promotes better performance in terms of activity and stability.

Much work has stipulated that using the Pt/ZrO2 catalyst and operating at relatively moderate temperatures with a $_{CO2/CH4}$ ratio equal to or greater than unity can minimize coke formation [56].

Furthermore, it has been shown that the hydroxyl groups existing on the surface of the interacting ZrO2 catalyst smell very strongly with $_{CO2\ and}$ generate bicarbonates and formates that are recognized as the main reaction intermediates in the dry methane reforming reaction [57].

Slagtern et al [59] and Bradford et al [56] have recommended, based on their work, the use of basic media to inhibit carbon deposition. Swaan et al. [60] and Hayakawa et al. [61] recommended using the carrier that could provide large dispersions of nickel to avoid carbon formation.

In addition, the literature has reported that CeO2 may be a suitable dopant or carrier for nickel [62]. It has been shown in recent years that CeO2 added to alumina significantly enhances catalytic performance. Thus, Wang [62] examined a series of Ni/Al2O3, Ni/CeO2 and Ni/Al2O3-CeO2 catalysts more closely. He found that Ni / Al2O3 - CeO2 shows better results in terms of catalytic activity and stability whereas, CeO2 alone (without alumina), used as a nickel support has a sufficiently strong interaction with the metal (SMSI strong metal support interaction) can significantly reduce catalytic performance.

Lercher et al [60] and Tsipouriari et al [63] propose to operate with supports with low concentration of Lewis acid sites and/or in the presence of basic sites such as ZrO2, MgO or La2O3 in order to improve catalytic performance and minimize deactivation by coking.

Schuurman et al [64] showed that the addition of MgO to a catalyst system based on Ru dispersed on inert carbon improves the performance of the base catalyst.

Wang et al [65] and Montoya et al [66] have developed a series of supported catalysts of different supports: MgO, Al2O3, TiO2 and CeO2. Contrary to what is reported, they found no difference in catalytic behaviour and no improvement in resistance to carbon deposition.

In the same context, Erdöhelyi et al [67] state that the activity of nickel-based catalysts depends mainly on the dispersion of the Ni particles but not on the support. This hypothesis is also confirmed by Slagtern [55] who assumes that only the metal/support interaction could control the dispersion and morphology of the nickel particles and thus the catalyst efficiency.

Erdöhelyi et al [67], do not note any difference between Rh based catalysts deposited on TiO2, Al2O3, SiO2, MgO as opposed to Pd, for which they observe the following sequence of decreasing activity: TiO2 > Al2O3 > SiO2 > MgO.

The use of zeolites as possible carriers for the dry reforming reaction has been discussed in many works. Thus, it has been shown that the introduction of transition metals into zeolites by ion exchange provides excellent catalytic predispositions [37].

The introduction of Rh particles into the matrix of a zeolite by the ion exchange method has been widely described in the literature. Bhat and Sachtler [37] examined Rh catalysts introduced into an NaY zeolite and obtained catalysts that are efficient in terms of activity, selectivity and stability.

Crisafulli et al [68] compared a zeolite support (ZSM-5) with a conventional support (SiO2) and found that when nickel is deposited on SiO2, it gives better results in terms of activity and stability. However, Ru deposited on SiO2 has a lower activity than when supported on ZSM-5.

V. DEACTIVATION OF DRY METHANE REFORMING CATALYSTS

The formation of carbonaceous deposits from the reagents can lead to surface fouling, blocking of the catalyst pores and/or disintegration of the support. Many authors, including Reitmeier et al [69], White et al [70], Sacco et al [71] have presented thermodynamic calculations to predict carbon deposit formation as a function of operating conditions. The main conclusion, following this study, is the need to work at more than 727°C with a CO_2/CH_4 ratio much higher than unity in order to avoid the thermodynamic zone of carbon formation. However, for the application of this process in the industrial field, it is recommended to work at a higher temperature than the unit in order to avoid the thermodynamic zone of carbon formation.

lower temperature with a co_2 / CH4 ratio close to the unit. It is mentioned in many works, that the inactive carbon formed during dry methane reforming, comes either from the methane decomposition reaction which is endothermic or the CO dismutation reaction which is exothermic. Calculations by Reitmeier et al [69] have shown that the equilibrium constant of the CO dismutation reaction decreases with increasing temperature, and conversely, the equilibrium constant of CH4 decomposition increases. The carbon formed during dry methane reforming is often in filamentary form [72]. Rodriguez stated in a review that the critical phase of carbon formation is the diffusion of carbon into metal particles [73]. The heat released by the Boudouard reaction would allow the diffusion of carbon.

Conversely, another study, based on kinetic data, states that the origin of carbon would probably be the decomposition of methane [71]. Moreover, Nolan et al [74] have insisted on the fact that carbon deposition is accelerated by the presence of H_2 in the reaction mixture.

In addition, the formation of carbonaceous deposits includes the production and transformation of several forms of carbon [75].

Koerts et al [76], Sacco et al [73], have identified three types of carbon on supported transition metals: $C\alpha$, $C\beta$, $C\gamma$. Among these types of carbon, $C\alpha$, which is an active species, would be responsible for the formation of CO while the less active $C\beta$, $C\gamma$ would be responsible for the deactivation of the catalyst.

Koerts et al [76] state that the $C\alpha$ also called carbidic carbon, can be hydrogenated at a temperature below 50°C and the $C\gamma$ or amorphous carbon, is reduced between 100°C and 300°C while the $C\gamma$ commonly called graphitic carbon is reduced at a temperature above 400°C.

However, no direct correlation has been established between the formation of $C\alpha$ and catalytic activity. However, other equally interesting correlations have been established by the team of Chen et al [77] for Ni-based catalysts deposited on Al2O3 and MgO. They found that the more active the catalyst is, the more carbon is formed. In the same study, Chen et al. observed the formation of two reduction peaks by H_2: one appeared between 227°C and 427°C which they attributed to the $C\alpha$ type carbon, the other around 600°C attributed to the $C\beta$ type carbon less reactive towards hydrogenation compared to the $C\alpha$ type carbon. It was concluded from this study that the two forms of carbon would play different roles in dry methane reforming.

The carbon $C\beta$ accumulated during the reaction would be responsible for the deactivation. These groups of researchers were able to show that the catalytic activity is proportional to the amount and reactivity of $C\alpha$. They found that the greater the dispersion of the catalyst, the less $C\beta$ is formed.

Another study has shown that the methane decomposition reaction and CO dismutation requires Ni atoms as active sites [78]. Other results showed that $_{CO2}$ would be at the origin of $C\alpha$ and that CH4 decomposition would be at the origin of $C\beta$.

We can see in these studies the difficulty of differentiating between carbonaceous species that actually participate in the catalytic cycle and species that, escaping from the cycle, gradually accumulate on the surface of the catalyst while evolving chemically towards more stable and therefore more toxic species.

B/ FRIEDEL-CRAFTS REACTION.

INTRODUCTION

Friedel-Crafts reactions, in particular alkylation reactions of aromatic compounds with aromatic halides lead to polyaromatic products. These compounds are used as special polymers in macromolecular chemistry as well as dielectric fluids in industry after removal of the catalyst and distillation of all condensation products.

In addition, Friedel-Crafts reactions are the basis of classical organic chemistry which allow the creation of C-C bonds. In the field of homogeneous catalysis, these reactions are catalysed by Lewis acids in the liquid phase such as: FeCl3, AlCl3 ... etc.

However, there are some problems with these catalysts [79], among others:

-Formation of stable complexes with the alkylation reagents or with the products, making them irrecoverable.

-Corrosion of installations, toxicity, pollution problems.

For this reason, some researchers have found it useful to test this type of reaction in the field of heterogeneous catalysis using catalytic systems that do not form stable complexes with the products, making it possible to recover them at the end of the reaction. Replacing liquid acids with solid materials will thus make it possible to reduce pollutant discharges. Indeed, the objective of this study is the development of a catalyst based on an efficient hydrotalcite for Friedel-Craft type reactions, at atmospheric pressure and at a reaction temperature not exceeding 80°C.

I. FRIEDEL-CRAFT REACTION

Alkylation and acylation are two reactions in which the reaction intermediate is the electrophilic (E+) species: a carbocation. The reaction mechanism results in a carbon-carbon bond. These reactions are often referred to as Fridel-Crafts reactions. Indeed we distinguish two types of reactions:

> **Alkylation**

It is a substitution of a hydrogen H with an alkyl group R on an aromatic ring. It also allows a side chain to be linked to a ring. It results from the reaction between a benzene hydrocarbon and an alkyl halide R-X in the presence of a catalyst. This reaction is very interesting as a method for the synthesis of side-chain benzene hydrocarbons. However, some disadvantages remain, namely :

-The difficulty of limiting the reaction to the substitution of a single hydrogen.

-alkyl does not always bond to the ring through the halogen bearing carbon in the alkyl halide R-X. We can cite the case of 1-chloropropane reacting mainly with benzene to give isopropylbenzene instead of propylbenzene.

-Ar-X aryl halides (e.g. chlorobenzene) are not suitable for this reaction. Therefore, two benzene rings cannot be welded together by this method.

> **Acylation**

It is a substitution of a hydrogen H with an acyl group R-CO; a ketone of the form Ar-CO-R is obtained. By analogy with alkylation , this product results from the reaction of a

Acid chloride (R-CO-Cl) on a benzene hydrocarbon in the presence of aluminium chloride promoting the formation of the carbocation R-C+=O (same principle as with alkyl halide) [80].

II. VALORISATION OF FRIEDEL-CRAFTS REACTIONS

Some applications of Fridel-Crafts' reactions in industry include the following:

a. Dielectric fluid synthesis

The dielectric fluid synthesis process comprises the following successive steps:

> ➤ Condensation of aromatic halides and aromatic compounds in the presence of catalysts
> ➤ Catalyst removal by neutralisation or washing
> ➤ Possible distillation of excess reagents.
> ➤ Distillation of all the condensation products or at least one of them followed by its conditioning in dielectric fluid. The improvement consists in removing the second step.

b. Synthesis of ethyl benzene

This synthesis takes place by alkylation of benzene with ethylene. This exothermic reaction requires *Lewis* acid catalysis. The oldest processes use aluminium trichloride AlCl3. The reaction takes place in the gas phase, at ^{1800C,} under a pressure of 9 atm.

c. Styrene synthesis

More than 90% of styrene production comes from the dehydrogenation of ethyl benzene. Moreover, by polymerization of this molecule, we obtain polystyrene which is used in all plastic products (plastic bags, yoghurt pots, packaging, etc.).

III. ALKYLATION REACTION CATALYSTS OF BENZENE BY BENZYL CHLORIDE

Figueras et al [81] have developed Montmorillionite K10 catalysts exchanged with transition metal ions. Tested in the alkylation reaction of benzene with benzyl chloride, they have led to the following main results:

> ➤ The study was carried out in the presence of ten different catalysts with the ratio (substrate/agent) not specified. They found that the conversion rate varies remarkably with the time of the benzylation reaction. The results show that some catalysts give a total conversion of benzyl chloride. On the other hand, the K-10-Ti (IV) catalyst has the best compromise between reactivity and selectivity.

> ➤ Unlike liquid catalysts used in homogeneous catalysis, good conversion of benzyl chloride is observed when using solid catalysts. Indeed, the catalytic efficiency increases respectively by 40% in the case of the liquid catalyst TiCl4 to 66% than in the case of the solid K-10-Ti (v).

> ➤ A low catalytic charge allows for better catalytic activity and good efficiency.

> ➤ The catalyst K-10-Fe (III) can be used without loss of activity by lengthening the reaction time in the benzylation by 8 times.

> ➤ The order of reactivity of any *Lewis* acid sequence for Friedel-Crafts reactions is as follows: **AlBr3, AlCl3> GaCl3 > FeCl3 > SbCl3 >** ZrCl4 **> BCl3 > SbF5.**

However, supported Zr (IV) has a better activity than supported Al (III) (concerning conversion and selectivity).

Choudhary et al [82,83] has also carried out several studies in this field. In some studies carried out on two types of catalysts: montmorillonites K10 and

Mesoporous Si-MCM-41 mesoporous silicates exchanged with transition metal chlorides. It has shown that catalytic performance depends greatly on the nature of the metal. The following activity classification was established:

FeCl3 > InCl3 > GaCl3 > ZnCl3.

In other studies, Choudhary has examined the influence of the support on the catalytic performance of solids [85,86]. He found that the support has a remarkable influence on catalytic performance and ranked them in descending order of performance as follows: **InCl3/ Mount-K10 > InCl3 / Si-MCM-41 > In2O3/H-ZSM-5.**

On the other hand, Choudhary et al [79, 87, 88] tested hydrotalcites as catalysts of Friedel-Craft reactions. They examined the following samples: (MgFe-HTc, Mg/Fe=3), (InMg-HTc, Mg/In=3), (GaMg-HTc, Mg/Ga=3). It was found that this type of material performs well in the benzylation reaction.

And even for the benzylation reactions of toluene and other aromatic compounds (anisole, xylene and naphthalenes).

BIOGRAPHICAL REFERENCES

F. Cavani, F. Trifirò, A. Vaccari, Catal. Today, 11 (1991) 173. [2]-
A. Vaccari, Catal. Today, 41 (1998) 53.

[3]-E. Álvarez- Ayuso, H.W. Nugteren, Wat Res, 99 (2001) 2534.

4]-D.P. Das, J. Das, K. Parida. J. Coll Int-Sci, 261(2003) 213.

[5]-U. Costantino, M. Curini, F. Montanari, M. Nocchetti, O. Rosati. J. Mol. Catal A:
Chem, 195 (2003) 245.

[6]-E. Álvarez- Ayuso, H.W. Nugteren, Chem, 62 (2006)155.

7]-O.P. Ferreira, S.G. de Moraes, N. Durán, L. Cornejo, O.L. Alves, Chem, 62 (2006) 80.

[8]-S. Ribet, D. Tichit, B.Coq, B.Ducourant, F. Morato, J.S. State. Chem, 142 (1999) 382.

[9]-H. Morioka, Y.Shimizu, M. Sukenobu, K. Ito, E. Tanabe, T. Shishido, K. Takehira.
Appl. Catal. A: Gen, 215 (2001) 11.

10]-A.C.C. Rodrigues, C.A. Henriques, J.L.F. Monteiro, Mat. Res, 4 (2003) 563.

11]-M.N. Bennani, D. Tichit, F. Figueras, J. chim Phys, 3 (1999) 498.

12]-M.R. Kang, H.M. Lim, S.C. Lee, S.H. Lee, K.J. Kim. J. Mat. Onl, 1 (2005) 1.

E. Kanezaki, Sol. Stat. Ion, 106 (1998) 279.

[14]-S. Casenave, H. Martinez, C. Guimon, A. Auroux, V. Hulea, A. Cordoneanu, E.
Dumtriu, Ther. Act 397 (2001) 85.

15]-L. Obalova, M. Valaskova, F. Kovanda, Z. Lacny, and K. Kolinova, Chem Pap. 58
(2004) 33.

16]-M. Marquevich, F. Medina, D. Montané, Catal Commun, 2 (2001) 119.

17]-D.M.Meira, G.G. Cortez, W.R. Monteiro, J.A.J.Rodrigues, J.Chem.Eng, 3 (2006) 351.

18]-V.J. Bulbule, V.H. Deshpande, S. Velu, A. Sudalai, S. Sivasankar, V.T.Sathe, Tetrah,
55 (1999) 9325.

[19]-A.I. Tsyganok, T. Tsunoda, S. Hamakawa, K. Suzuki, K. Takehira, T Hayakawa. J.
Catal, 213 (2003) 191.

20]-T.Shishido, M.Sukenobu, H.Morioka, R.Furukawa, H.Shirahase, K.Takehira, Catal
Lett, 73 (2001) 21.

[21]-Z. ming Ni, G.X. Pan, L.G. Wang, W.H.Yu, C.P. Fang, D. Li. J. Chem. Phys, 19
(2006) 3.

Mr. Sigel. M.C.J Bradrof, H. Knözinger, M.A. Vannice, Top Catal, 8 (1999) 211.

23]-N.R. Udengaard, J.H.B. Hansen, D.C. Hanson , J.A. Stal , *Oil & Gas , 90* (1992) 62.

[24]-A. Erdöhely, J. Cserényi, E. Papp , F. Solymosi, Appl. Catal. A Gen, 108 (1994) 205.

[25]-S. Teuner, Hydrocarbon Processing, 64 (1985) 106.

26]-C.Tsang, J.B. Claridge, M.L.H. Green, Catal. Today, 23 (1995) 3.

27]-J.R. Rostrup-Nielsen , J.H. Bak-Hansen, J. Catal, 144 (1993) 38.

[28]-P.D.F. Vernon, M.L.H. Green, A. K. Cheetam , A.T. Ashcroft, Catal. Today, 13 (1992) 417.

[29]-A.T. Ashcroft, A. K. Cheetam, M.L.H. Green, P.D.F. Vernon, Nature, 352 (1991) 225.

30]-V.C.H. Kroll, H.M. Swaan, C. Mirodatos, J. Catal, 161 (1996) 409. 31]- Z.L. Zhang, X.E. Verykios, J. Chem. Soc., Chem. Commun. 71 (1996).

[32]-E. Ruckenstein, Y.H. Hu, Appl. Catal. A: Gen, 133 (1995).

[33]-E. Ruckenstein, H.Y. Wang, Appl. Catal. A: Gen, 204 (2000) 257.

[34]-J.H. Sinfelt, Bimetallic Catalysts: Discovery, Concepts and Applications, (Wiley, New York, 1983).

[35]-C. Crisafulli, S. Scirè, R. Maggiore, S. Minicò , S. Galvagno, Catalonia, Latvia, 59 (1999) 21. [36]-D. Halliche, R. Bouarab, O.Cherifi, M.M. Bettahar, Catal. Today, 29 (1996) 373.

[37]-R.N. Bhat, W.M.H. Sachtler, Appl. Catal. A, 150 (1997) 279.

[38]-D. Halliche, O.Cherifi, A. Auroux, J.Therm.Anal. Cal. 68 (2002) 997.

39]-J.S. Chang, S.E. Park, K.W. Lee, M.J. Choi, Stud. Surf. Sc. 84 (1994) 1587.

40]-Zhanlin Xu, Ming Zhen, Yingli Bi , Kaiji Zhen, Catal. Lett, 64 (2000) 157.

[41]-N. Lyi, S. Takekawa , S. Kimura, J. Solid State Chem, 83 (1989) 8.

42]-O.S.W. Perz-Lopez, A. Senger, N.R. Marcilio, M.A. Lansarin, App.Catal A: 303 (2006) 234.

[43]-A. Battacharyya, V.W. Chang, D.J. Schumacher, Appl Catal Sci, 13 (1998) 317.

[44]-Z. Hou, T. Yashima. Apll Catal A: Gen 261 (2004) 205.

45]-J.H. Edwards, A.M. Maitra, Fuel. Technol. Proc. 41 (1995) 269.

[46]-Y.Y. Wu, O. Kawaguchi, T. Matsuda, Bull. Chem, Soc.Jpn., 71 (1998) 563.

47]-T.Hayakawa, S. Suzuki, J. Nakamura, T. Uchijima, S. Hayakawa, K. Suzuki, T. Shishido, K. Takehira, Appl. Catal. AGen, 183 (1999) 273.

[48]-A. Battacharyya, V.W. Chang, Stud.Surf.Sci. Catal. 88 (1994).

49]-J. Guo, H. Lou, H. Zhao, D. Chai, X. Zheng, Appl. Catal. A Gen.,273 (2004) 75.

50]-P. Ferreira-Aparicio, A. Guerrero-Ruiz, I. Rodríguez-Ramos, Appl. Catal. A: Gen. 170 (1998) 177.

[51]-T. Uchijima, J. Nakamura, K. Saito, K. Aikawa, K. Kubushiro, K. Kunimori, Elsevier Sci. B.V. Natural Gaz Conversion, 2 (1994) 81.

52]-J. Nakamura, K. Aikawa, K. Sato, T. Uchijima, Catal. Lett, 25 (1994) 265.

53]-V.A. Tsipouriari, A.M. Efstathiou, Z.L. Zhang, X.E. Verykios, Catal.Today, 21(1994) 579.

54]-Z.L. Zhang, V.A. Tsipouriari, A.M. Efstathiou, J. Catal, 158 (1996) 51.

55]-M.C.J. Bradford, M.A. Vannice, J.Catal., 173 (1989) 17.

56]-S.M. Stagg, E.Romeo, C.Padro , D.E. Resasco, J. Catal, 178 (1998) 137.

57]-J.H. Better, K. Seshan, J.A. Lercher, J. Catal, 171 (1997) 279.

58]-J.A. Lercher, J.H. Better, W. Hally, W. Niessen, K. Seshan, Stud. Surf. Catalan Science, 101 (1996) 463.

[59]-A. Slagtern, Y. Schuurman, C. Leclerq, X.E. Verykios, C. Mirodatos, J.Catal.,172 (1997) 118

60]-H.M. Swaan, V.C.H. Kroll, G.A. Martin , C. Mirodatos , Catal. Today, 21 (1994) 571.

[61]-T. Hayakawa, H. Harihara, A.G. Andersen, A.P.E.York, K. Suzuki, K. Takehira, Angew. Chem. Int. ed. 35 (1996) 192.

[62]-S.Wang, G.Q. Lu, Appl. Catal. B, 19 (1998) 267.

63]-V.A. Tsipouriari, X.E. Verykios, J.Catal., 187 (1999) 85.

64]-Y.Schuurman, C.Mirodatos, P. Ferreira-Aparicio, I. Rodríguez-Ramos, A. Guerrero-Ruiz, Catal. Lett, 66 (2000) 33.

[65]-S.Wang, G.Q. Lu, Appl. Catal. A, 169 (1998) 271.

66]-J.A. Montoya, E. Romero-Pascual, C. Gimon, P. Del Angel, A. Monzon, Catal. Catal. Today, 63 (2000) 71.

[67]-A. Erdohelyi, J. Cserényi, F. Solymosi, J. Catal, 141 (1993) 287.

[68]-C. Crisafulli, S. Scirè, S. Minicò , L. Solarino, , Appl. Catal. A Gen, 225 (2002) 1.

69]-R.E. Reitmeier, K. Atwood , H.A. Bennet, J.r. and H.M. Baugh, Ind. Eng. Chem.,40 (1948) 620.

70]-G.A. White, T.R. Roszkowski, D.W. Stanbridge, Hydrocarbon Process, 54 (1975) 130.

[71]-A. Sacco, Jr, F.W.A.H. Geurts, G.A. Jablonski, S.Lee , R.A. Gatell, J. Catal, 119 (1989) 322.

[72]-J.T. Richardson, S.A. Paripatyadar, Catalan Application, 61 (1990) 293.

[73]-N. M. Rodiguez, J. Mater. Res, 8 (1993) 3233.

74]-P.E. Nolan, D.C. Lynch , A.H. Cutler, Carbon , 32 (1994) 477.

[75]-C. Bartholomew, Catal. Rev. 24 (1) (1982) 67.

[76]-T. Koerts, R.A. Van Santen, J. Chem. Soc., Chem. Commun, (1991) 1281.

77]-Y.Chen, J. Ren, Catal. Lett, 29 (1994) 39.

P. Ferreira-Aparicio, I. Rodríguez-Ramos, A. Guerrero-Ruiz, , Appl. Catal. A: Gen, 148 (1997) 343.

79]-V.Choudhary, R. Jha, P.Choudhary, J. Chem.Sci 6 (2005) 635.

80]-G.A. Olah, Friedel-Crafts Chemistry, Wiley, New York, 1973.

F. Figuears, T.Cseri, S. Bekassy, S. Rizner, J.Mol. Catal. A: Chem.98 (1995) 101. 82]-V.R. Choudhary, S.K. Jana, J. Mol. Catal.A, 180 (2002) 267.

83]-V.R. Choudhary, S.K. Jana, B.P. Kiran, Catal.Lett. 64 (2002) 233.

84]-V.R. Choudhary, S.K. Jana, M.K. Chaudhari, J. Mol. Catal.A, 170 (2002) 251.

85]-V.R. Choudhary, S.K. Jana, Appl. Catal. A, 244 (2002) 51.

86]-V.R. Choudhary, R. Jha, V.S. Narkhede, J.Mol. Catal A: Chem 239 (2005) 76.

87]-V.R. Choudhary, S.K. Jana, V.S. Narkhede, Appl.Catal A: Gen 235 (2002) 207.

Chapter II
Techniques experimental

CHAPTER II

CUSTOMARY METHODS FOR CHARACTERIZING LAMELLAR DOUBLE HYDROXIDES

INTRODUCTION

In the field of catalysis, we cannot simply look at the catalyst only in its reactive aspect, so a good understanding of the catalytic act often requires in-depth studies of the intrinsic physico-chemical properties that could allow possible correlations between catalytic behaviour and some of the characteristics of the sample.

In the case of hydrotalcites, various techniques are generally used as shown in Figure-1 [1].

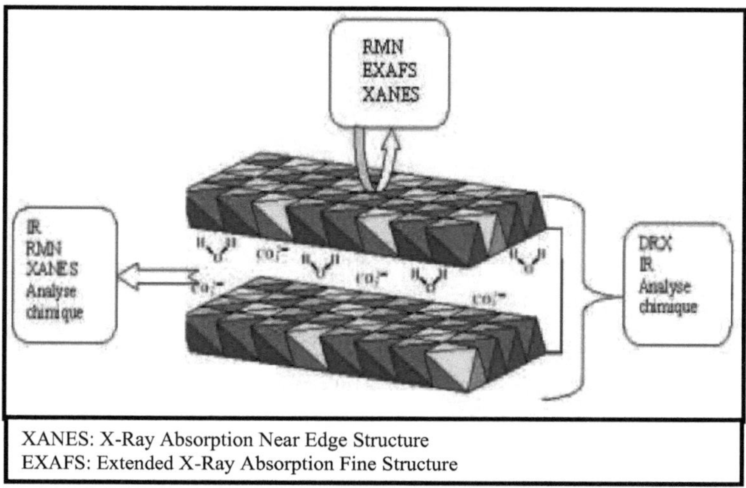

XANES: X-Ray Absorption Near Edge Structure
EXAFS: Extended X-Ray Absorption Fine Structure

Figure-1: Different techniques used for the characterization of Hydrotalcites.

In this chapter we will describe the different techniques used such as: X-ray diffraction (XRD), specific surface area measurement (BET), chemical analysis (ICP), infrared spectroscopy (FTIR), and scanning electron microscopy (SEM).

I. CHEMICAL TEST

Atomic absorption is one of the most widely used methods in the elemental chemical analysis of catalytic systems. Advantages include sensitivity, speed of execution, selectivity of the method, among others.

Elemental chemical analysis allows us to propose a chemical formulation of the material and to confirm the so important ratio M (II)/ M (III).

The chemical analysis of our samples was carried out on a SOLAAR969AAA SPECTOMETER type apparatus whose schematic is shown in Figure-2.

The content of the element under consideration is given by the formula :

$$A\ (\%) = \frac{0.25.\ \text{Concentration}}{PE}.100$$

A (%): percentage of the element considered i.

PE: Test sample = 0.50g.

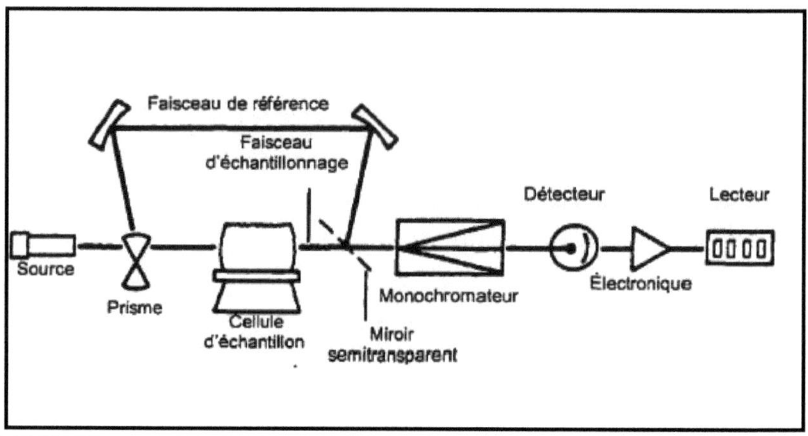

Figure-2: Simplified diagram of the double beam atomic absorption spectrometer

II. X-RAY DIFFRACTION

The use of this method is intended to determine - among other things - the degree of crystallinity of the samples, as well as the crystallographic structure. It is based on Bragg's law:

$$n \lambda = 2 d \sin\theta$$

Where:
n: integer corresponding to the order of diffraction. λ:
wavelength of the radiation used ($\lambda_{K\alpha}$=1.54 A°). d: inter-
reticular distance (A°).
θ: diffraction angle.

The radiocrystallography study of our catalysts was carried out on an X' PERT PRO MPD type diffractometer, the operation mode of which is shown in Figure-3.

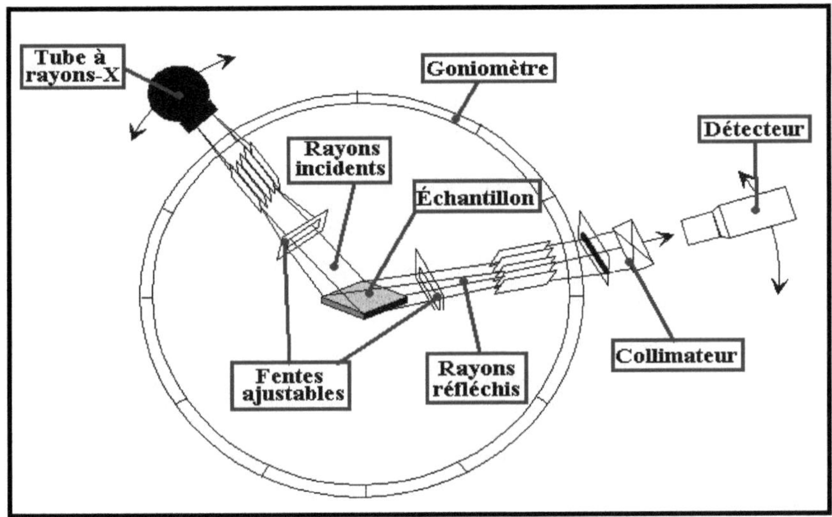

Figure-3: Schematic diagram of an X-ray diffraction spectrometer.

III. FTIR INFRA RED SPECTROSCOPY This is probably the most widely used spectroscopic technique, since all polyatomic molecules have an infrared spectrum.

This makes it easy to obtain information on the surface properties of the sample.

To do this, the sample is pelleted in KBr and placed on a sample holder and inserted into the instrument for analysis.

The analysis by infrared spectroscopy is carried out here using a PERKIN-Elmer spectrometer coupled to a computer for recording, storing and processing the spectra.

VI. SCANNING ELECTRON MICROSCOPY (SEM)

The morphology and crystal size of our materials were obtained by scanning electron microscopy (SEM). The studies are done on a PHILLIPS- XL type apparatus.

V. THE BET METHOD (Brunauer-Emmet-Teller)

The catalytic reaction is a reaction that occurs on the surface of the solid. The catalytic activity of a solid is often given per mole/unit of surface area. The knowledge of the available surface likely to be in contact with the reagents is thus an essential data for the understanding of the catalytic act.

The isotherm established by *BRUANAUER, EMETT and TAYLOR* based on physisorption [2], is used to evaluate the surface of a solid (commonly referred to as the specific surface). The technique is based on the determination of the volume required to form a monolayer (Vm) and the surface area of the molecular surface of the adsorbed molecule. The kinetic method leads to the following equation:

$$\frac{(P/P0)}{v_a(1 - P/P0)} = \frac{1}{Vm.C} + \frac{(C-1)}{Vm.} \frac{P}{P}$$

P: equilibrium adsorption pressure.

P0: saturating vapour pressure of the adsorbent at the temperature of the experiment.

C: constant which depends on the adsorption and liquefaction heat of the gas.

v_a: volume of gas adsorbed at pressure P.

Vm: volume of gas required to establish a complete monomolecular layer.

The BET linear transformation is plotted by plotting **(P/ P0)/ v_a (1- P/P0) as a** function of **P/P0.** with: 0.05< P/P0 < 0.35.

We obtain a straight line of slope **C-1 /Vm** C (with C-1 / C close to 1) and the intercept at the origin is 1 /Vm **C.**

The specific surface area **(S) of** the catalyst is given by the following relationship, reduced to one gram of the catalyst.

$$S = \sigma \; N \; Vm \; / \; 22414$$

Knowing Vm, we can determine **S**

with:

      σ : the area of the average cross section of the adsorbed nitrogen molecule (16 A°2)

      N: Avogadro number.

      Vm: expressed in cc/g.

      S: expressed in m2/g.

DESCRIPTION OF THE CATALYTIC TEST

The reaction of methane reforming with carbon dioxide is carried out in the gas phase at atmospheric pressure in a fixed-bed differential dynamic reactor with a sintered glass on which a thin layer of powdered catalyst is arranged to avoid temperature gradients. The temperature of the reaction is determined by means of a thermocouple arranged at the level of the catalytic bed in the furnace. The arrangement of the catalytic test is shown in Figure 4.

> ➢ The gaseous reactants (CH_4 and $_{CO_2}$, Ar) are preheated at the reactor inlet and outlet by means of heating cables. The CH_4, $_{CO_2}$ and Ar gas circuits are all equipped with a pressure gauge and a shut-off valve. A Brooks flowmeter provides consistent and accurate flow rates.

> ➢ The analysis of reagents (CH_4, $_{CO_2}$) and products ($_{H_2}$, CO) is carried out regularly by periodic injections (every 30 min) by an *IGC121ML* gas chromatograph (GC). Retention times and integrated peak areas are given by a ***Schimadzu*** computer integrator ***(CR8A) CHROMATOPAC.***

> ➢ A water trap (with 0°C ice) is placed at the reactor outlet to retain water vapour before any chromatographic analysis.

➢ The flow measurement after reaction is carried out by a soap bubble flowmeter at the outlet of the chromatograph.

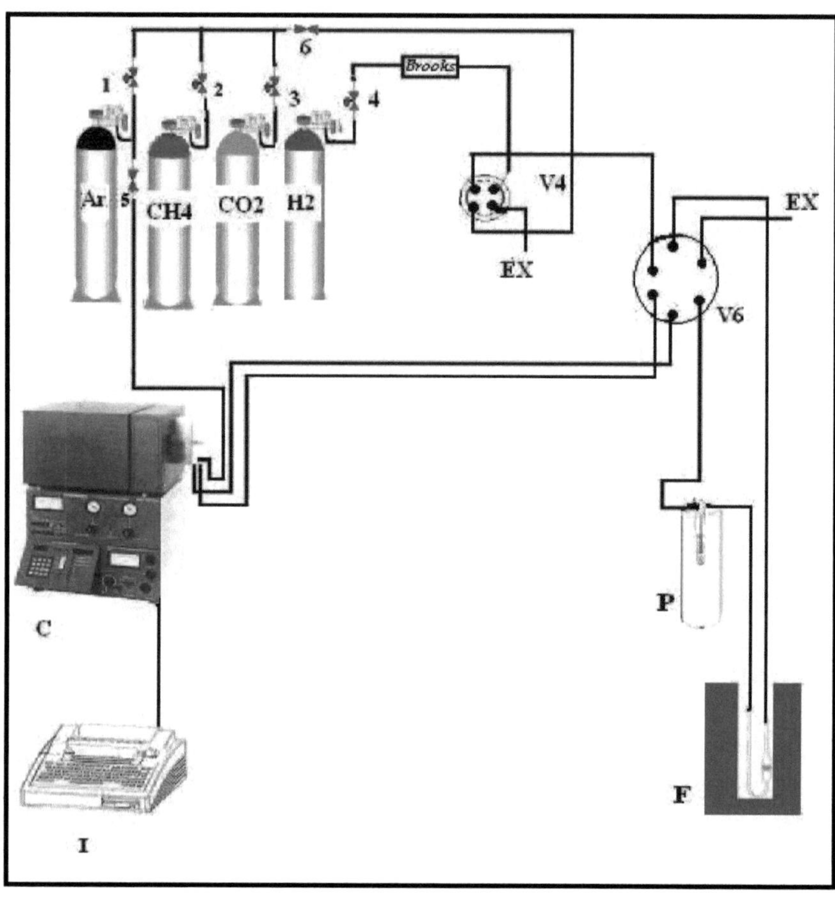

Figure-4: Experimental set-up of the catalytic test.

I: Integrator	P: Water trap	C: Chromatograph
V6: Six-way valve	A: Reactor	F: Oven
V4: Four-way valve	5, 6: Micro valves	EX: Way outdoors
1, 2, 3 and 4: Shut-off valves		

THE ELEMENTS OF THE CATALYTIC TEST

The catalytic test consists of several elements including :

1. FOUR

The oven is a vertical cylinder. It is equipped with a carbolite-type temperature regulator. It makes it possible to obtain the high temperatures necessary for the dry methane reforming reaction. The temperature of the reaction is determined by a thermocouple located at the catalytic bed in the furnace.

2. REACTOR

The reactor is a U-shaped quartz tube, vertically arranged inside the furnace, with a sinter that carries the catalytic charge (Figure-5).
One end of the reactor allows the reaction mixture to enter, the other end returns it to the chromatographic analysis by passing it through a water trap.

3. WATER TROLLEY

It's a glass tube immersed in a dewar vase containing ice. It's used to condense water vapour as it exits the reactor. It is shown in Figure 6.

4. DEBIMETRE BROOKS

A flow controller type BROOKS (*MODEL 5878*) is designed to regulate the hydrogen flow of the reduction.

5. SOAP BUBBLE DEBIMETRE

It is used to measure the passage time of a given volume of gas.

6. HEATING CABLES

Their role is to maintain the ducts at a sufficiently high temperature to avoid gas condensation.

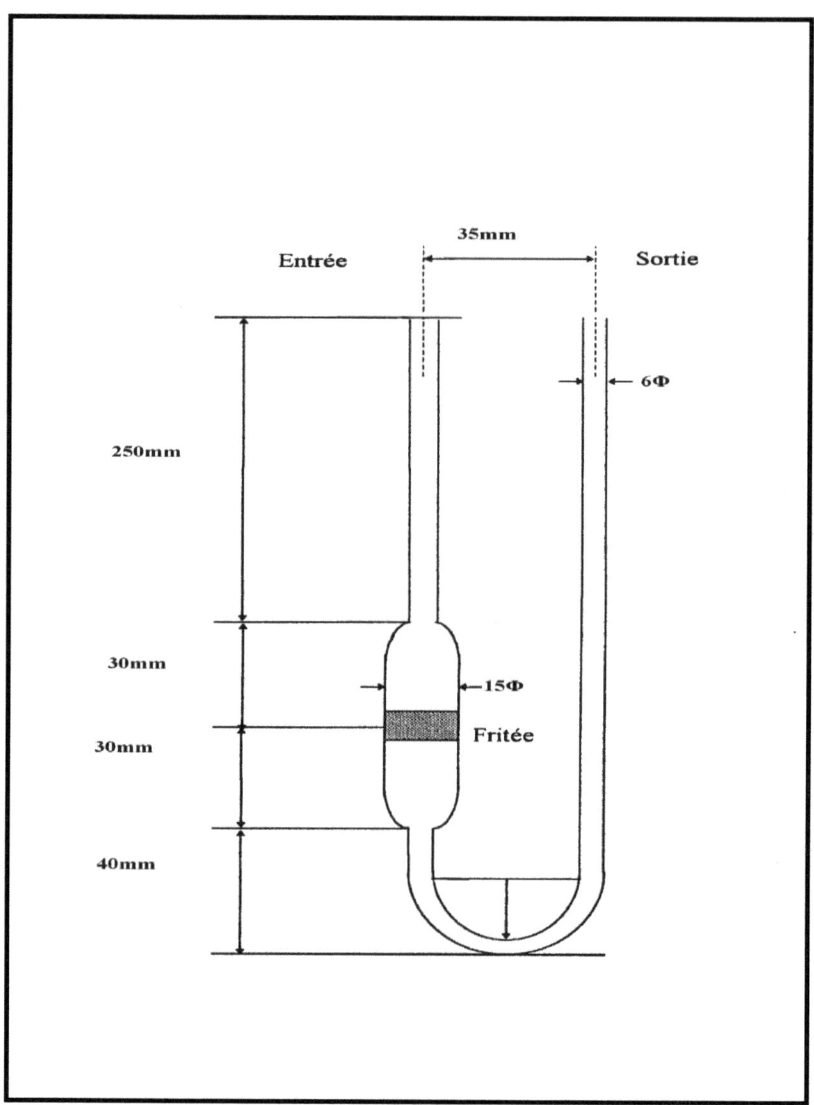

Figure-5: Fixed Catalytic Bed Quartz Reactor.

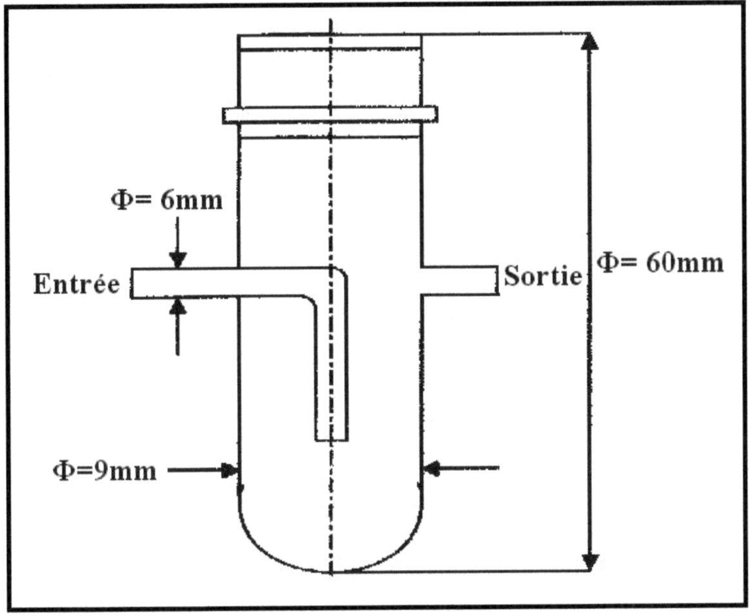

Figure-6: Diagram of the water trap.

7. FOUR-WAY VALVE (V4)

It allows either the reduction gas or the reaction mixture to be sent to the reactor (Figure-7).

8. SIX-WAY VALVE (V6) :

This valve allows the orientation of the reaction mixture either towards the analytical chromatograph and then the reactor, or directly towards the reactor and then the chromatograph (Figure 7).

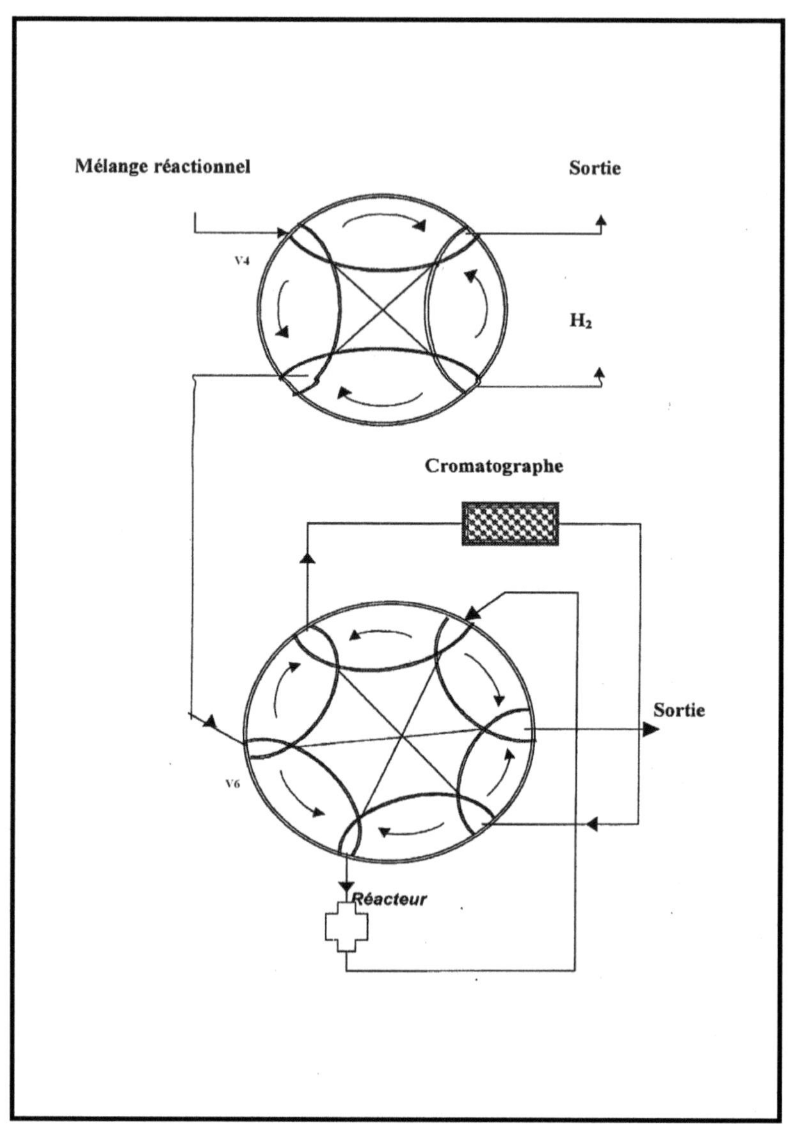

Figure-7: Operation of valves V6 and V4.

ANALYTICAL CHROMATOGRAPHIC METHOD

The analysis of the gas mixture before and after the reaction is carried out by gas chromatography.

Product separation is carried out on a two-metre long stainless steel column containing a carboseive phase.

A chromatograph type *IGC121ML* (TCD) is used. The carrier gas is argon. All experiments were performed at atmospheric pressure.
The operating analytical conditions of the chromatograph are [3]:

Oven temperature   *= 100°C.*
Injector temperature = 100°C Detector

temperature = 100°C
Accessories temperature = 70°C

Qualitative analysis

 Reagents and reaction products are identified by comparing retention times with those of pre-calibrated products as shown in Table 1.

Table-1: Retention time of products and reagents according to calibration.

Standard mixture	CH4	CO	CO2	H2
Retention time	3,30	1 ,70	7,20	1,20

We give an example of a chromatogram (Figure-8) obtained by injecting a gas mixture of known composition.

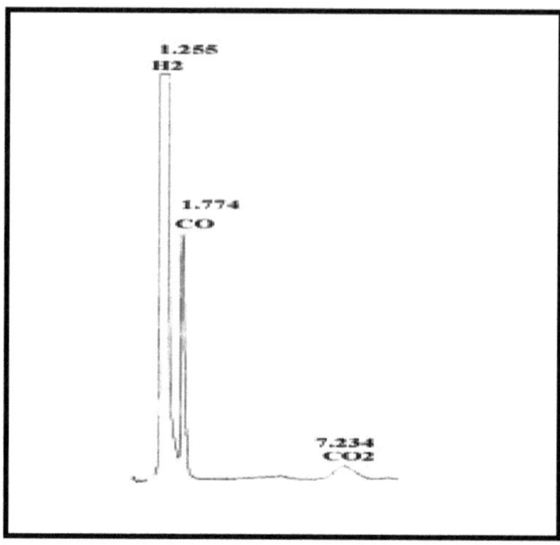

Figure-8: Chromatogram of the standard mixture.

Quantitative analysis

The quantitative analysis of the products and reagents is carried out using the response coefficients determined beforehand by a calibration (Eq. 1).

The response factor **Kfi** is, by definition, the response of the chromatograph *to* compound 'i'. We calculate it by the following relationship:

$$\mathbf{Kfi = (Xi\,/Si)\,.^{104}} \qquad (Eq.1)$$

Where:

Xi: molar percentage of compound "i" in the standard mixture. si: the peak area of compound "i" given by the integration.

The response factor for each gas in the standard mixture is given in Table 2.

Table-2: Composition of the standard mixture and response factors.

Mixture and standard	CH4	CO	CO2	H2
Composition (%)	10,00	15,40	5,98	78,67
Retention time	3,92	1,73	7,22	1,21
Response factor	0,68	1,73	2,43	0,20

Before each catalytic test, "blank" tests were carried out under the same experimental conditions by "bypassing" the reactor. As a comparison, two examples of chromatograms are given. One is given after the blank test (Figure-10-A), and the other after passing through the reactor (Figure-9-B).

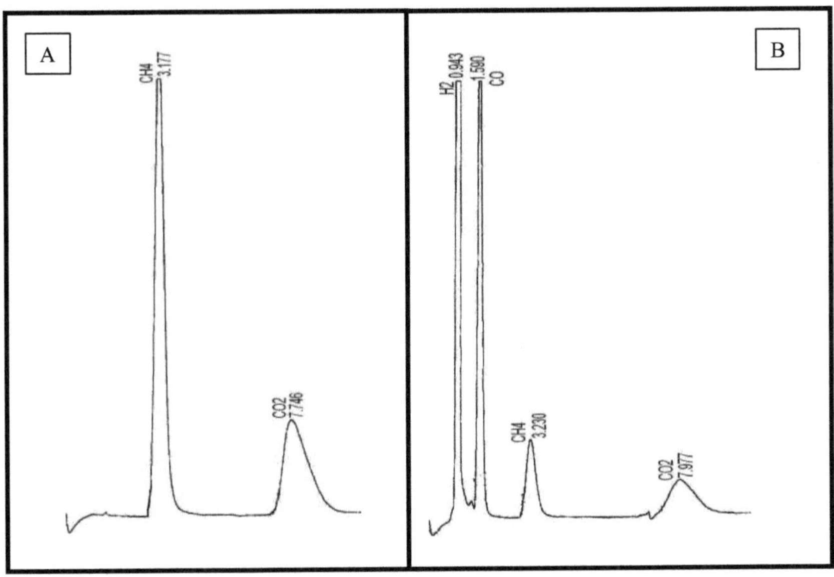

Figure-9: Examples of chromatograms of the reaction mixture before catalytic test (A) and after catalytic test (B).

TESTCATALYTIC KNOTTING

a. Catalyst activation

The catalyst activation step is very important for its reactivity. The activation of the catalyst is carried out by reduction under hydrogen at a rate of 1.2 L/h at atmospheric pressure for 12 hours.

b. Catalytic reaction

After insitu reduction of the catalyst, the furnace temperature is maintained at the reaction temperature, and the hydrogen gas of the reduction is replaced by the reaction mixture (CH_4, CO_2) by means of the four-way valve (V4). The flow rate and composition of the reaction mixture are adjusted beforehand and the reagents and products are sent through the six-way valve (V6) to the chromatograph for analysis.

c. Estimation of reagent compositions and products

The volume compositions of the flue gases are calculated according to the following expression :

$$\boxed{Xi\,(\%) = S_i.\,Kfi\,.10^{-4}} \qquad (Eq.2)$$

With:

Xi: Composition of compound (i).

I_f: Area of the peak corresponding to the compound (i), given by the integrator.

Kfi: Response factor of the compound (i).

d. Estimation of catalytic activity

The comparison of catalytic performances is carried out by calculating CH_4 and CO_2 (Eq. 3) conversions, which we note here TTGi, as well as synthesis gas productivity or n CO (Eq.4). Carbon balances are estimated via (Eq.5).

$$\boxed{\begin{array}{l} {}^{From}.\ Xi^{\,e} - Ds\,.Xis \\ TTGi(\%) = \text{------------------}\ Xi^{\,e-} \\ Ds\,.Xis\ TTGi(\%) = \text{---------------} \\ Xi^{\,e-}Ds\,.Xis\ TTGi(\%) \end{array}} \qquad (Eq.3)$$

e. Estimation of CO productivity (nCO)

Carbon monoxide productivity is estimated by the following relationship:

$$nCO \text{ (mol/h.g)} = \frac{XSCO \cdot DS}{22.4 \cdot mcat} \qquad \text{(Eq.4)}$$

With:

nCO: Data in mole/h.g.

XSCO: Composition of carbon monoxide.

mcat: Mass of the sample (0.1g).

f. Estimation of theoretical carbonaceous product balances

This balance allows to give the order of magnitude of coke deposit formed during the CH4 / CO_2 reaction, this carbon loss is estimated by the following expression :

$$100 \cdot XC \,(\%) = \frac{Ds. \,(XSCH4 + XSCO2 + XSCO)}{From \cdot (Xe_{CH4} + Xe_{CO2})} \qquad \text{(Eq.5)}$$

With:

DS: Effluent flow rate at the reactor outlet.

From: Effluent flow at reactor inlet.

XSCH4, XSCO2, XSCO: Volume composition at the reactor outlet.

XeCH4 ,XeCO2 : Volume composition at the reactor inlet.

g. Estimation of apparent activation energies

The apparent activation energy is determined by *Arrhenuis'* law:

$$\text{TTGi} = \text{A EXP} (- \text{Eai} / \text{RT})$$

TTGi: Rate of disappearance of reagent i (i: CO_2, CH4).

A: *Arrhenuis* constant. **A:**

Perfect gas constant.

Eai: Apparent activation energy of reagent i.

T: Reaction temperature (°C).

This energy is calculated from the line: **Ln (TTGi)** as a function of **(1 /T).**

$$\text{Ln TTGi} = \text{Ln A} - \text{Eai} / \text{RT}$$

Ln A is deduced by extrapolating the line at the origin of the ordinate axis and the activation energy is calculated either from the slope or from the expression, after calculation of the *Arrhenuis* constant.

h. Estimation of synthesis gas selectivity

Syngas selectivity is estimated by the ratio H_2/ **CO.**

THE ALKYLATION REACTION OF BENZENE WITH BENZYL CHLORIDE

a. Equipment

The alkylation reaction of benzene with benzyl chloride is carried out, at atmospheric pressure, in a tri-glued flask. The reaction mixture is heated to the reaction temperature and the catalyst is introduced into the reactor. We then note the start time t0. Every 20 minutes we withdraw 1µl from the solution with a syringe. The solution is analysed by a chromatograph (at the CRAPC-USTHB research centre).

The alkylation reaction of benzene with benzyl chloride is as follows:

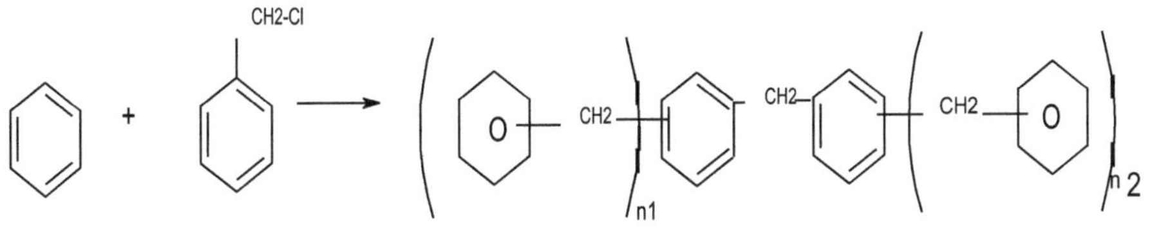

Where: n1 and n2 are 0, 1 or 2 and n1+ n2 is less than or equal to 3.

Benzene + Benzyl Chloride $\longrightarrow$

♠ Diphenyl methane (n1+ n2 = 0).

♠ Dibenzyl benzene (n1+ n2= 1).

♠ Tribenzyl benzene (n1+ n2= 2).

An internal standard was used to correct calculation errors. In our case, we use pentadecane since its boiling temperature is close to that of benzyl chloride.

The molar ratio used in the catalytic tests is: Benzene to

benzyl chloride = 15.0.

Benzyl chloride / pentadecane = 2.0.

b. Apparatus for analysing the reaction mixture by gas chromatography

We have adopted gas chromatography to separate and analyze our gas mixtures. For this we have used :

- A **G.C.H.P 6890** chromatograph, equipped with a **FID** type detector, operating isothermally and having as carrier gas nitrogen.
- The column used is of **HP5MS** type, non-polar with low selectivity, this phase tends to separate the compounds according to their boiling point (temperature: 50 to 325°C).
- A typical Enhanced Chemstation G170 BA Version B.01.00, Copyright-© Hewelett-Packed 1989-1998 software that presents and calculates the peak areas of gaseous compounds.

c. Chromatographic calibration

1. Analytical condition of the FID chromatograph

The operating conditions are shown in Table-3.

Table-3: Chromatographic Analysis Conditions.

Capillary column	- HP5MS. - Non-polar. - Separates products according to their boiling point.
Initial column temperature	140 °C
Final column temperature	280°C
Detector temperature	300°C
Initial time	0 min
Final time	20 minutes

2. Qualitative analysis

➤ Determination of the retention times of the analyzed gases

The reagents and reaction products shall be identified by comparison of the retention times with those of the pre-calibrated products.

Table 4 reports the retention times tR of the reaction products and reagents obtained on FID.

Table-4: Retention times for products and reagents (FID).

Reagents	benzene (BZ)	Benzyl chloride (ClBZ)	Pentadecane C15	Diphenyl methane (DPM)
tR (min)	2,14	2,67	2,43	6,5

3. Quantitative analysis

➤ Calculation of the response factors Kfi of the analyzed gases

The response factor of a compound "i" is the chromatograph response (FID in our case) to that compound. The response factor for each element in our reaction mixtures and products (Table 5) was determined by performing calibrations using standard bottles.

Table-5: Response Factor Values of Reagents and Products from Calibration.

Element	Kfi
BZ	720247
ClBZ	955815
C15	822249
DPM	1413825

4. Implementation of the catalytic test

> **Catalyst activation**

Catalyst activation consists of calcination under air (2 L /h) at atmospheric pressure for 3 hours at 300°C (5°C/ min). The purpose of this pre-treatment is to desorb the water and carbonates that would have been adsorbed on the surface of the catalyst.

5. Calculation formulas used :

> **Estimation of the conversion of benzyl chloride**

The conversion of benzyl chloride is estimated by the following formula:

$$\textbf{Conv}_{ClBz} = (\textbf{S0ClBz} - \textbf{S}_{f\,ClBz}) / (\textbf{S0}_{ClBz}).$$

With:

$$S0ClBz\ (corrected) = (s_{ClBz})_{0/}$$
$$(Standard).\ Sf_{ClBz}\ (corrected) =$$
$$(s_{ClBz})/(Standard).$$

> **Estimation of selectivity**

The selectivity for diphenyl carbon methane is estimated by the following law:

$$\textbf{DPM selectivity} = ([c]_{DPM}\,\textbf{formed}) / \Delta\,[c]_{ClBz}\,\textbf{disappeared})$$

With:

$$\Delta\,[C]_{disappeared} = [C]\ \text{in mole of initial ClBz} - [C]\ \text{in mole of final ClBz}.$$

$$[C]ClBz = \text{Scorched ClBz}/\,Kf_{ClBz}$$

$$C]ClDPM = DPM_{Scoring} /\,Kf_{DPM\,Scoring}$$

$$S_{correctedClBz} = s_{ClBz} \times (Standard)_{0} / (Standard)_{f}$$

$$Scaled_{DPM} = SDPM \times (Standard)_{0/} (Standard)_{f}$$

BIOGRAPHICAL REFERENCES

F. Cavani, F. Trifirò, A. Vaccari, Catal. Today, 11 (1991) 173.

[2]-G. Leofanti, M. Padovan, G. Tozzola, B. Venturelli. Catal. Today 41 (1998) 207. [3]-A. Boudjemaa. Magister's thesis "Catalytic production of hydrogen from the gas-to-water conversion reaction. Application in photocatalysis on FeCr2O4", (U.S.T.H.B, 2006).

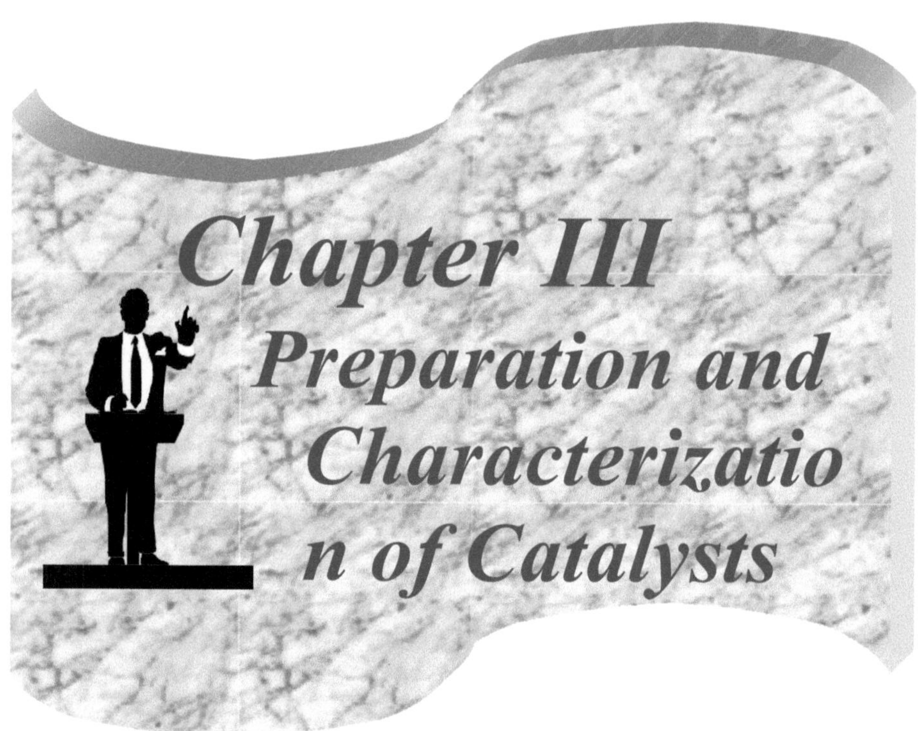
Chapter III
Preparation and
Characterizatio
n of Catalysts

CHAPTER III

INTRODUCTION

The catalyst generally results from a complex sequence of unitary steps in which the texture, structure, mass and surface composition and oxidation state of the metals are likely to evolve in an uncontrolled manner. It is therefore important to set the operating parameters for each unit stage in the preparation of catalytic systems.

Our materials are lamellar double hydroxides of the cations M^{2+} and M^{3+}. The structure is based on a stack of lamellar sheets of composition $M(OH)_2$, similar to those of brucite $Mg(OH)_2$ 1. The substitution of the divalent Mg^{2+} ions by other divalent or trivalent cations is conditioned by the value of the ion radii of the M^{2+} and M^{3+} ions (which must be close to 0.65 A°, the value corresponding to the Mg^{2+} ions) 1. The electrical neutrality of the structure is ensured by an interleaf space containing anionic entities solvated by water molecules 1,2,3. The general formula of the hydrotalcite structure is :

$$\boxed{\; M1{-}x^{2+}\, M^{3+}\,(OH) \qquad [^{An-}]_{x/n} \cdot m\; H2O. \;}$$

I. CATALYST PREPARATION

1. CHOICE OF PREPARATION METHOD

All of our catalyst systems have been developed using the co-precipitation method. This method consists of simultaneously precipitating at least two elements in a generally aqueous solution.

2. DESCRIPTION

Our catalysts are synthesized by the co-precipitation method based on a solution of (A) salts of cations M(II) and M(III) and a solution (B) of sodium bicarbonates. This method was chosen for its reproducibility and ease of performance. Moreover, this method of development allows to obtain double lamellar materials of basic character [1-3].

Reagents

> ***Solution A***

Nitrate salt of a divalent metal: M (II) (NO3)$_2$. 6H2O.

Nitrate salt of a trivalent metal: M (III) (NO3)$_3$. 9H2O.

Table 1 lists the different solutions of cations used in the preparation of catalysts.

Table-1: Products making up Solution A.

The salts in solution A	Chemical formulations of salts	Origin	Purity (%)
Nickel Nitrates	*Ni (NO3)$_2$. 6H2O*	*Merck*	*99,00*
Magnesium nitrates	*Mg (NO3)$_2$. 6H2O*	*Merck*	*99,00*
Aluminium nitrates	*Al (NO3)$_3$. 9H2O*	*Fluka*	*98,00*
Cobalt Nitrates	*Co (NO3)$_2$. 6H2O*	*Aldrich*	*97,00*
Lanthanum nitrates	*The (NO3)$_3$. 9H2O*	*Aldrich*	*99,90*
Iron Nitrates	*Fe (NO3)$_3$. 9H2O*	*Merck*	*97,00*
Crome Nitrates	*Cr (NO3)$_2$. 6H2O*	*Merck*	*98,00*
Copper Nitrates	*Cu (NO3)$_2$. 6H2O*	*Merck*	*99,00*

> ***Solution B***

-Sodium bicarbonates: Na2CO3.

-Sodium hydroxides: NaOH.

The same alkaline solution B was used for all catalysts. The characteristics of alkaline solution B are shown in Table 2.

Table-2: Products used to make up alkaline solution B.

	Chemical Formula	Origin	Purity (%)
Sodium hydroxide	NaOH	Merck	99,00
Sodium bicarbonate	Na2CO3	Merck	99,50

Method of operation

The method of preparation consists of adding the alkaline solution B, drop by drop, to the cationic salt solution A, using a pump so as to maintain the pH at a constant basic value. The mixture is kept under vigorous agitation until a precipitate in the form of a gel is obtained (Figure-1).

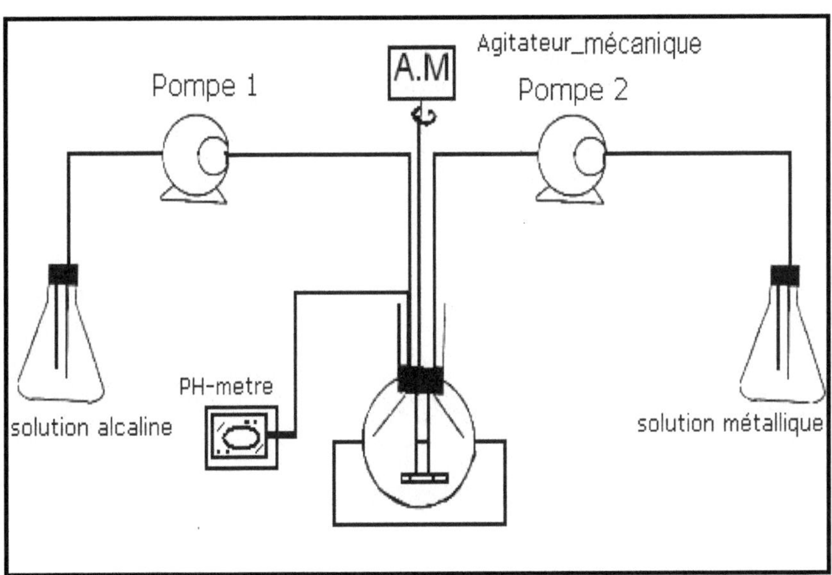

Figure-1: Assembly used in the preparation of hydrotalcite.

Washing

The resulting gel is washed several times with doubly distilled water to remove impurities.

Drying

This operation is carried out in an oven, at a temperature of 80°C.

The calcination

This operation is carried out in a tubular furnace whose temperatures can reach up to 1000°C. Each sample is calcined at 450°C under air for 6 hours at a heating rate of 4°C/min.

The formatting

After calcination, the catalyst must be shaped. For this purpose, the material obtained after calcination is crushed and then sieved so that it has a grain size of 0.16 mm in diameter.

The nomenclature

Samples developed on the basis of salts of cations M(II) (where M(II) = Ni, Cu, Cr Co, Mg) and M(III) (where M(III) = Al, Fe or La) are noted as follows:

- The non-calcined catalyst: NiMgAl-HT, NiAl-HT, NiMgLa-HT, NiLa-HT, CoMgAl-HT and finally CoMgLa-HT etc...
- Calcined catalysts: NiMgAl-HTc, NiAl-HTc, NiMgLa-HTc, NiLa-HTc, CoMgAl-HTc and CoMgLa-HTc etc...

II. CATALYST CHARACTERIZATION

1. CHEMICAL ANALYSIS AND SPECIFIC SURFACE

The chemical analysis of calcined samples is determined by atomic absorption, after attack of the solid by HNO3. This method of analysis allows us to determine the content of each element constituting the catalyst, which could lead us to a well-defined chemical formulation of the catalytic system. The results of the chemical analysis are shown in Table 3.

Table-3: Chemical analysis and chemical formulas of synthesized catalysts.

Catalysts	X	M2+/ M3+	Chemical Formulas	Specific Surface Area (BET) (m2/g)
NiMgAl-HTc	0,33	1,97	Ni0.2Mg0.45Al0.33	69
NiAl-HTc	0,29	2,40	Ni0.70Al0.29	68
CoMgAl-HTc	0,31	2,20	Co0.14Mg0.54Al0.31	---
NiLa	0,30	2,30	Ni0.69La 0.30	61
NiMgLa	0,31	2,20	Ni0.13Mg0.55La0.31	---
CoMgLa	0,31	2,20	Co0.11Mg0.57La0.31	---

It can be seen from this table that the experimental ratio (M2+/ M3+ = 2) is close to the theoretical ratio expected at the outset. On the other hand, the values of x obtained are between 0.29 and 0.33 corresponding to the optimal values of a good crystallization of the hydrotalcite structure as demonstrated in the literature in many works [4,5].

Table-3 also groups the values for specific areas. We note that the specific surface areas of our solids after calcination are of the same order of magnitude.

2. X-RAY DIFFRACTION

The radiocrystallographic analysis of the samples was carried out in a vertical powder goniometer with the aim of highlighting the different phases likely to make up our materials. This technique was used in the presence of both calcined and uncalcined samples.

a/ Uncalcined samples

The X-ray diffractograms of the various uncalcined samples are shown in Figures-2-5.
The X-ray diffraction spectra obtained showed average crystallinity of the lamellar double hydroxides. In the case of this type of uncalcined material, two types of important main peaks allow the identification of the double lamellar structure of hydrotalcite. These peaks are indeed found on the spectra of the samples :

-Intense, sharp, symmetrical peaks: (003), (006), (110) and (113).
-Asymmetrical wide peaks   : (012), (015) and (018).

These two types of peaks constitute a reference or imprint of the framework of typical hydrotalcite also called "brucite". Uncalcined hydrotalcites are also considered to be "isomorphic compounds" of brucite, since brucite could be obtained by a simple replacement of the Mg^{2+} and Al^{3+} cations by other divalent and trivalent cations in the brucite matrix, such as Ni^{2+}, Co^{2+}, Cu^{2+}, Cr^{2+} and/or La^{3+}, Fe^{3+} [6,7].

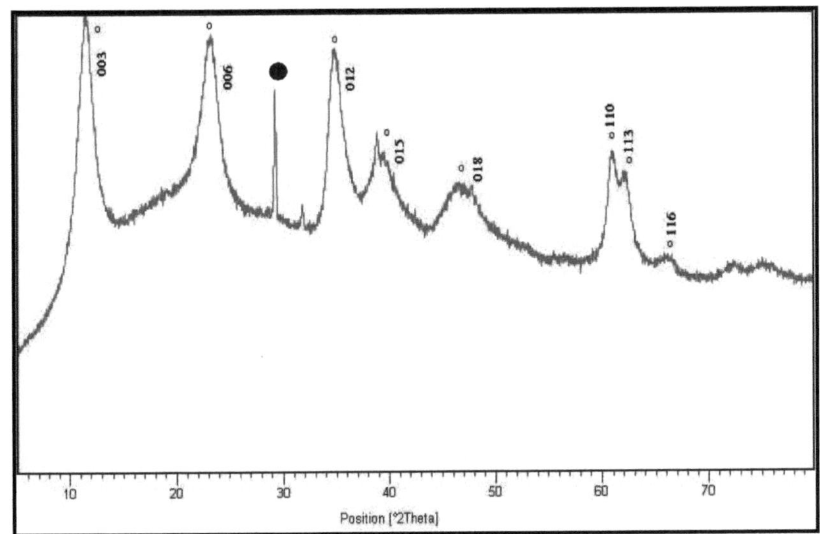

Figure-2: X-ray diffractogram of the uncalcined NiMgAl-HT sample. Hydrotalcite phase. ● Al(OH)₃.

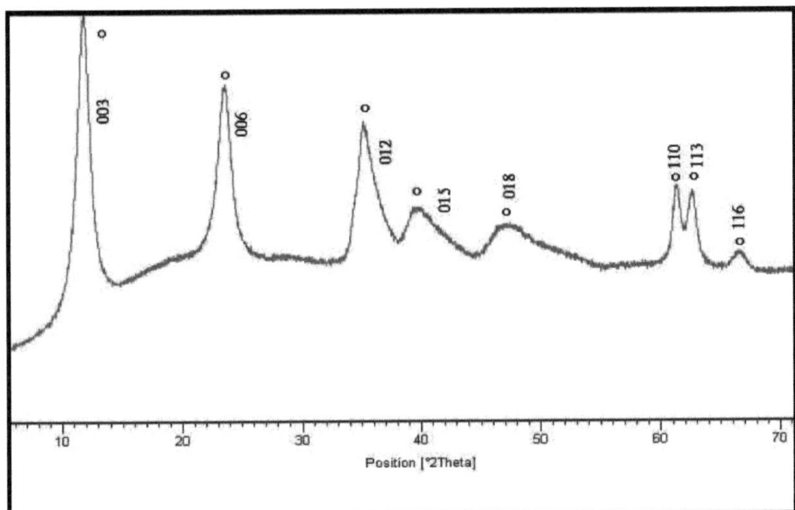

Figure-3: X-ray diffractogram of the uncalcined NiAl-HT sample. Hydrotalcite phase.

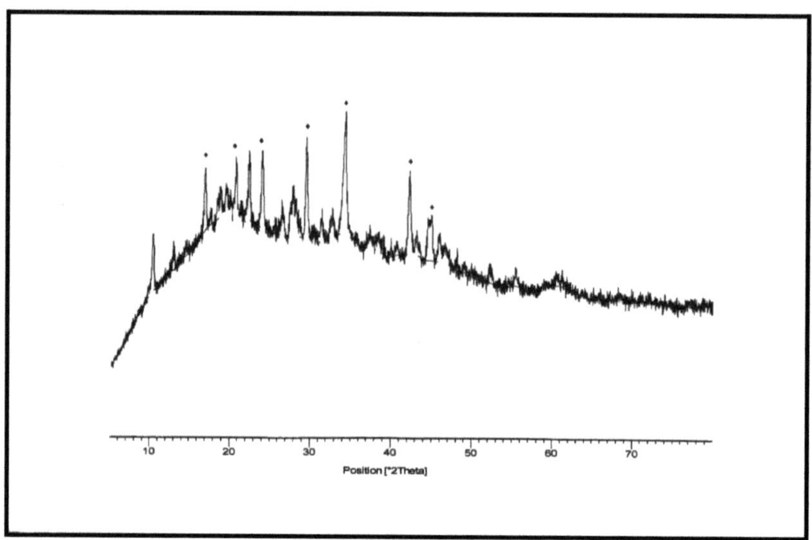

Figure-4: X-ray diffractogram of the uncalcined NiLa sample.
♦ M (CO3)$_x$ (OH)$_m$ y $_{H2O}$ (M = Ni, La).

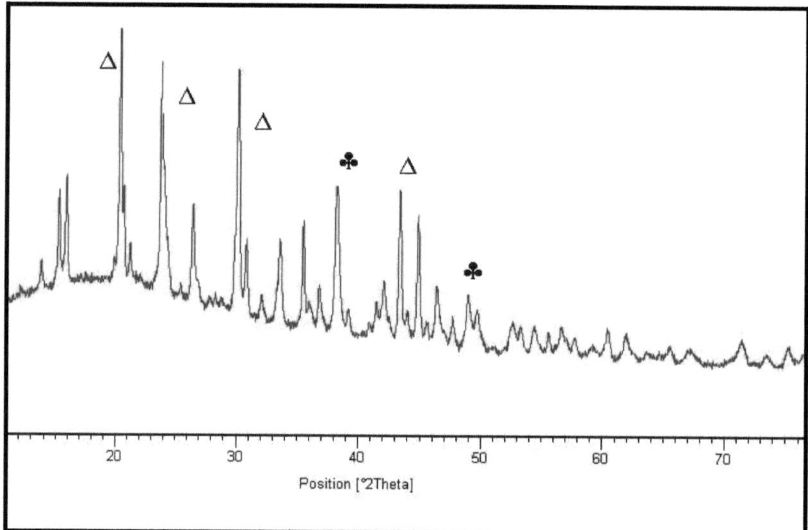

Figure-5: X-ray diffractogram of the uncalcined CoMgLa sample.
M(CO3)$_x$(OH)$_m$ y $_{H2O}$, (M = Co, Mg, La), Mg(OH)$_2$.

We can thus confirm from this study, that the structure of hydrotalcite is obtained in the majority of cases except for the samples at La, we suppose that the large size of the radius La3+ made that the structure is not obtained as shown in Figure-4 indeed, the obtaining of the typical structure of hydrotalcite is nevertheless limited by the nature of the cations M(II) and M(III) and their size. The most frequent M(III) cations which satisfy the double lamellar structure would be those whose ionic radius varies between 0.50 A° (Al) and 0.69 A° (Cr), and for M(II) cations the ionic radius varies between 0.65A° (Mg) and 0.80A° (Mn) [8,9].

b/ Calcined samples

The literature has reported at length on the great influence of heat treatment and its impact on the double lamellar structure [1, 2, 10, 11]. We present here some examples of X-ray diffraction spectra recorded for some samples after calcination (Figures-6-9). The literature has shown that the disappearance of the structure of hydrotalcites begins when the temperature exceeds 200°C [12].

Calcination of hydrotalcite leads to dehydration and dehydroxylation followed by decarboxylation in the case where the interleaf anions are carbonates. This last operation is accompanied by the collapse of the lamellar structure followed by the reduction of the interlamellar distance.

Numerous studies have reported the evolution of the double lamellar structure as a function of heat treatment [13].

We summarize them in the following series of equations:

$$Mg_{1-x}\,Al_x\,(OH)_2\,(CO_3)_{x/2}.m\,H_2O \xrightarrow{25\text{-}200°C} Mg_{1-x}\,Al_x\,(OH)_2\,(CO_3)_{x/2} \xrightarrow{225\text{-}500°C}$$

$$x/2\ MgAl_2O_4 + (1-x/2)\ MgO.$$

Moreover, Kanezaki [13], proposed a decomposition step for hydrotalcites :

$$Mg_{1-x}\,Al_x\,(OH)_2\,(CO_3)_{x/2}.m\,H_2O \xrightarrow{25\text{-}200°C} Mg_{1-x}\,Al_x\,(OH)_{2+x}.yH_2O \xrightarrow{225\text{-}500°C}$$

$$x/2\ MgAl_2O_4 + (1-3x/2)\ MgO.$$

Where: $y = m - x/2$.

This explains why the phenomenon of decarboxylation takes place in the first place according to the following diagram:

$$CO_3^{2-} + H_2O \longrightarrow CO_2 + 2\,OH^-.$$

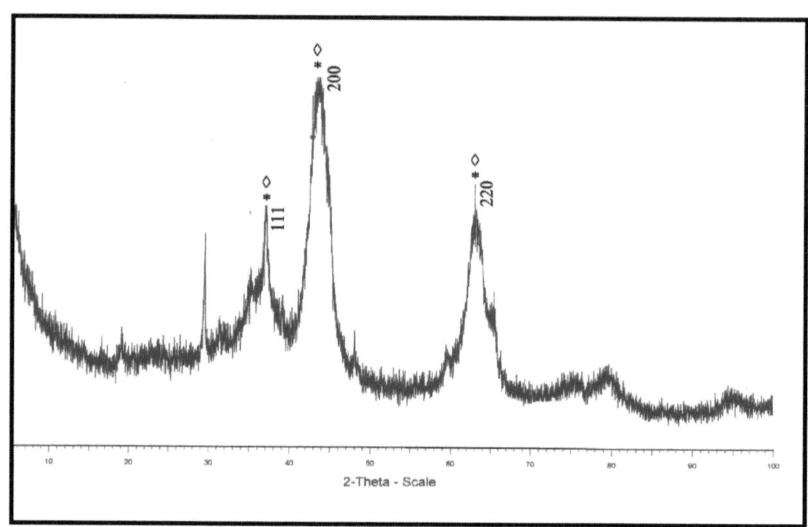

Figure-6: X-ray diffractogram of the NiMgAl-HTc sample calcined at 450°C.
Mgo, NiO

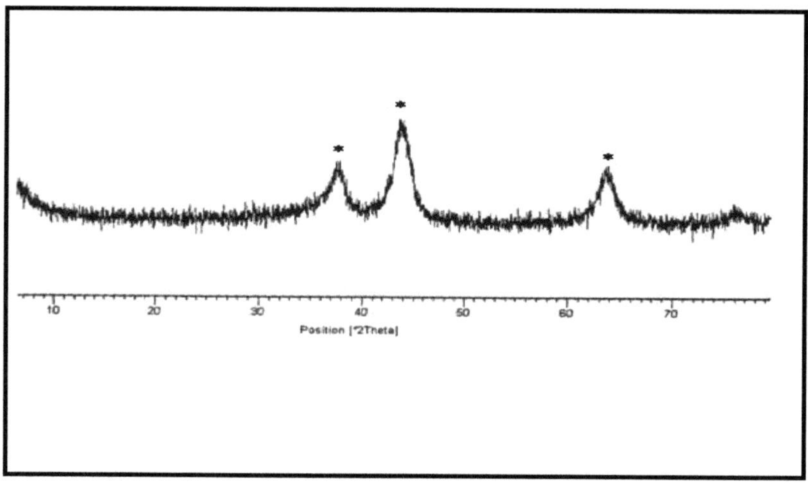

Figure-7: X-ray diffractogram of the NiAl-HTc sample calcined at 450°C. **NiO**

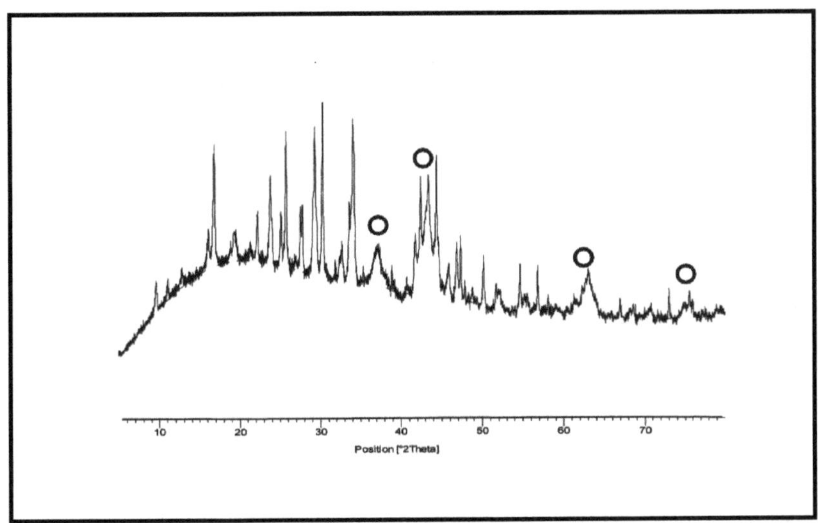

Figure-8: X-ray diffractogram of the NiLa sample calcined at 450°C.
o NiO.

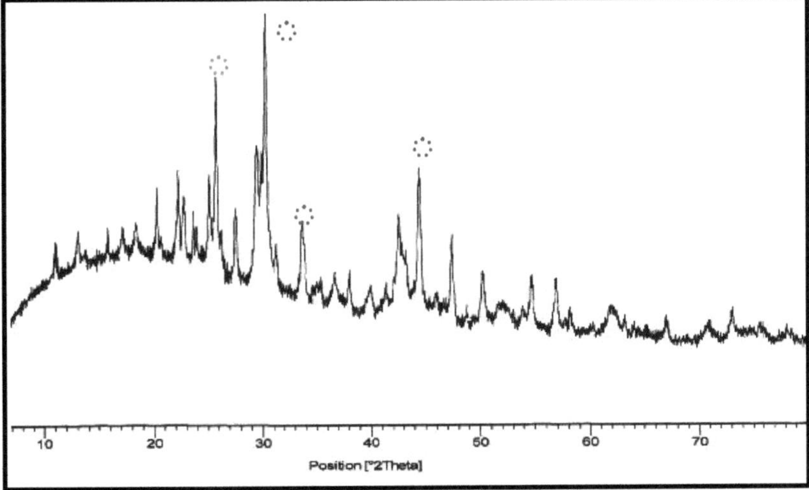

Figure-9: X-ray diffractogram of the calcined CoMgLa sample at 450°C. La2O3,
MgO,   Co3O4.

3. INFRARED SPECTROSCOPY (FTIR)

It is an analytical method that allows the identification of chemical species present on the surface of catalysts, as well as the understanding of the phenomena and reaction mechanisms that govern the transformation of these species.

Thus, our samples are examined by infrared spectroscopy in a vibration range from 400 to 4000 cm-1 (with a resolution of 2 cm-1).

a) Uncalcined samples

As examples, infrared spectra of NiMgAl-HT, NiAl-HT and CoMgAl-HT samples and their MgLa counterparts are shown in Figures 10 and 11, respectively.

In general, we note that the spectra reveal the same type of vibration bands over all the samples examined, i.e. a wide vibration band in the vicinity of 3300-3600 cm-1 which we can attribute to the hydroxide groups (OH) of the hydrotalcite framework and to the intercalated and/or physisorbed water molecules.

The vibration band observed in the vicinity of 1623-1630cm-1 could characterize the presence of intercalated hydroxyl groups of water molecules. In addition, the vibration band observed around 1350-1400 cm-1, would be associated with carbonates CO_3^{2-} of D3h symmetry (plane symmetry) as reported in the literature [8,9,14].

Furthermore, the vibration bands observed at low values (1000 cm-1) are characteristic of the M-O vibration mode [4,5,14].

b) Calcined samples

The heat treatment of our materials leads to the partial or total modification (depending on the temperature and the nature of the heat treatment) of the hydrotalcite structure as shown by the DRX analysis of the calcined samples. The analysis by infrared spectroscopy of the calcined samples also reveals these same modifications. Thus, after calcination at 450°C, dehydration and decarboxylation essentially occurs, which could be partial or total as shown in Figures -12 and 13.

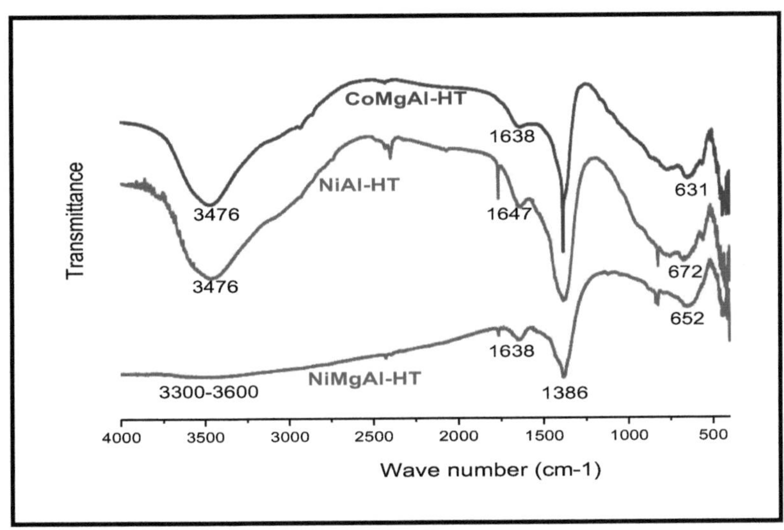

Figure-10: Infrared spectra of MgAl- hydrotalcite samples uncalcined HT

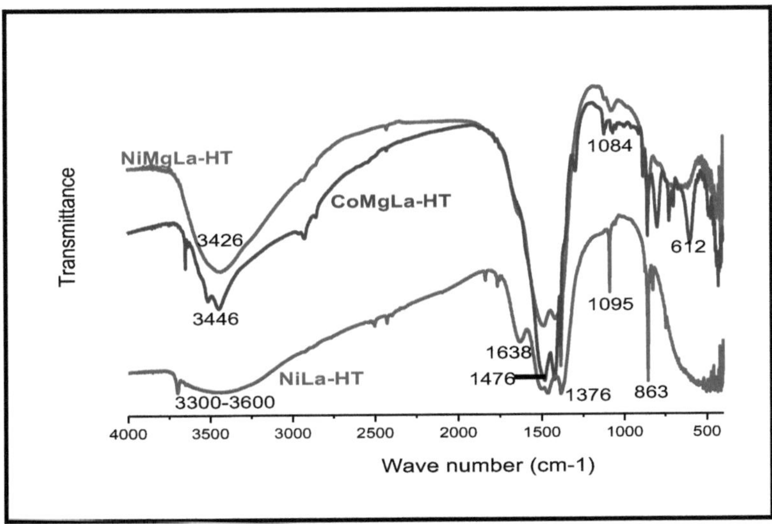

Figure-11: Infrared spectra of uncalcined MgLa series samples

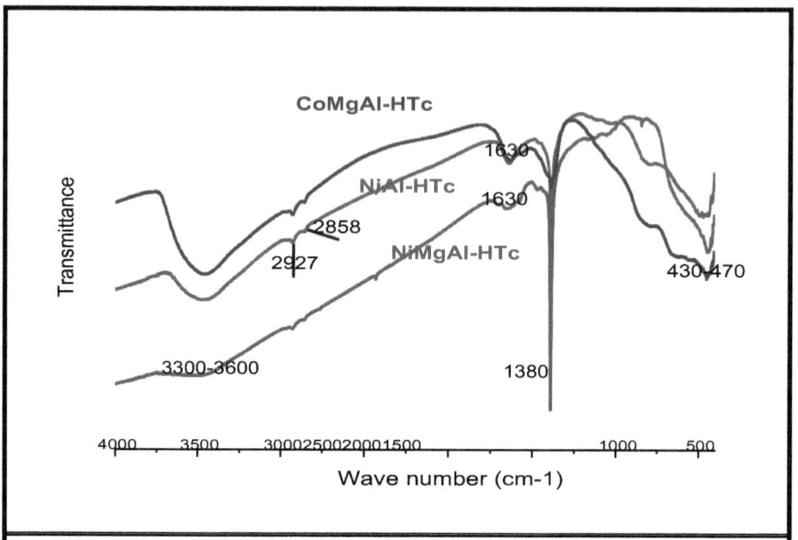

Figure-12: Infrared spectra of MgAl- HTc hydrotalcite-based samples calcined at 450°C

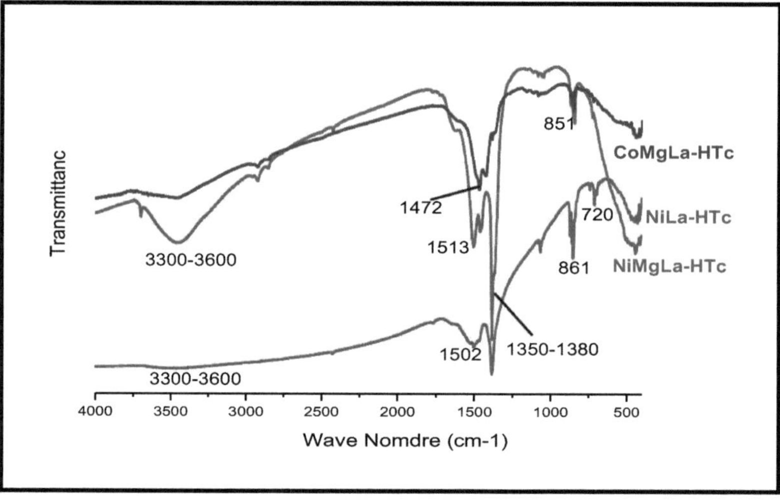

Figure-13: Infrared spectra of MgLa series samples calcined at 450°C.

As shown in these figures, the results obtained in this study for calcined and uncalcined samples appear to be similar and consistent with the results in the literature [1,14].

After calcination, the spectra essentially show a decrease in the intensity of the bands previously observed in the vicinity of 3500 cm-1, 1600 cm-1, and 1400 cm-1. Dehydration and decarboxylation would be at the origin of this decrease and even the disappearance of certain bands. On the other hand, the characteristic vibration bands of persistent carbonates where decarboxylation is only partial. As shown in the literature, the residual carbonate species graft themselves to the hydrotalcite framework, thus preventing the total collapse of the structure as shown in Figure-14.

This phenomenon gives hydrotalcites what we call the "memory effect", which, thanks to residual carbonates and rehydration, would allow the double lamellar structure of the original hydrotalcite to be recovered [2,3,10,15].

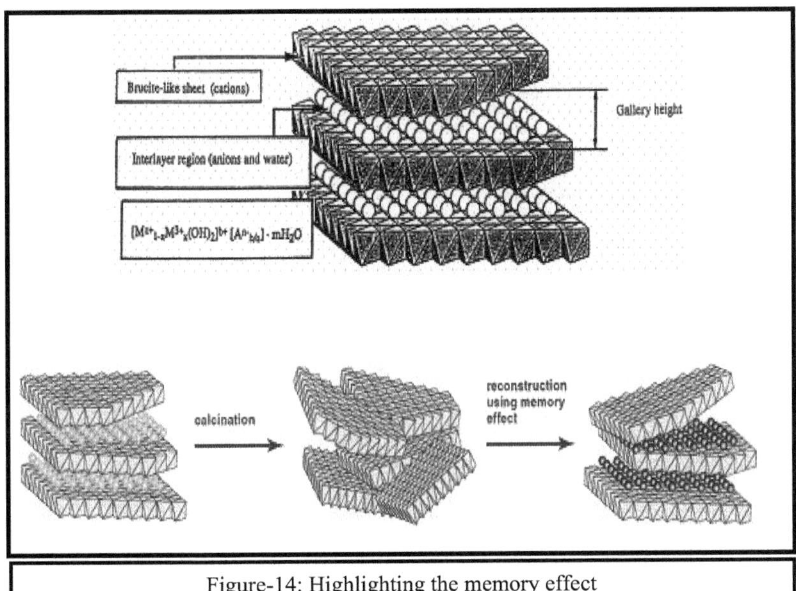

Figure-14: Highlighting the memory effect

4. FLUORESCENCE X

X-ray fluorescence is a technique mainly known in the field of quantitative analysis, but it is also a method that provides detailed information on the electronic structure of elements. On the other hand, this technique can complement the results obtained by X-ray diffraction analysis.

Table 4 shows the results obtained after analysis of the samples calcined at 450°C by X-ray fluorescence.

Table 4: X-Ray Fluorescence Results

Catalysts	NiMgAl-HTc	NiAl-HTc	NiMgLa	NiLa	CoMgAl	CoMgLa
Phases identified	NiO, MgO, Al2O3	NiO, Al2O3	NiO, MgO, La2O3	NiO, La2O3	Al2O3, Co3O4, MgO	La2O3, MgO, Co3O4

We clearly notice the presence of oxides of divalent and trivalent metals as previously shown by the X-ray diffraction technique.

CONCLUSION

- We have prepared a series of hydrotalcite-type catalysts by the co-precipitation method at basic pH based on bivalent and trivalent cations Ni, Co, Cr, Cu, Mg, La and Al. The overall chemical formula of these solids is :
$[M^{2+}_{1-x} M^{3+}_{x} (OH)_2]^{x+} [A^{n-}_{x/n}] . m H_2O$.

- These samples were examined by different physico-chemical analytical techniques, i.e. atomic absorption, X-ray diffraction, infrared spectroscopy, fluorescence.

- Chemical analysis by atomic absorption clearly confirmed that the molar ratio M(II)/M(III) is close to the initially expected value (value close to 2).

- Examination of the samples prior to calcination by X-ray diffraction confirmed that the hydrotalcite structure was obtained in the various cases.

- After calcination of the samples, the X-ray diffraction spectra reveal the formation of oxide phases and decrease or almost disappearance of the hydrotalcite phase.

- The analysis of the MgAl-HTc samples by infrared spectroscopy confirmed the hydrotalcite structure. It also revealed the presence of the same types of vibration bands in all cases.

BIOGRAPHICAL REFERENCES

[1]-E. Álvarez- Ayuso, H.W. Nugteren, Chem, 62 (2006)155.

F. Cavani, F. Trifirò, A. Vaccari, Catal. Today, 11 (1991) 173. [3]-

A. Vaccari, Catal. Today, 41 (1998) 53.

4]-A.C.C. Rodrigues, C.A. Henriques, J.L.F. Monteiro, Mat. Res, 4 (2003) 563.

5]-O.P. Ferreira, S.G. de Moraes, N. Durán, L. Cornejo, O.L. Alves, Chem, 62 (2006) 80. [6]-U. Costantino, M. Curini, F. Montanari, M. Nocchetti, O. Rosati. J. Mol. Catal A: Chem, 195 (2003) 245.

7]-V.J. Bulbule, V.H. Deshpande, S. Velu, A. Sudalai, S. Sivasankar, V.T.Sathe, Tetrah, 55 (1999) 9325.

8]-Z.Abdelssadek, F.Touhra, K.Bachari, A.Saadi, O.Cherifi, D.Halliche, Adv micro mesop mat, Heron press, Sofia (2008).

9]-J.A. Rivera, G. Fetter, Y.Jiménez, M.M.Xochipa, P. Bosch, Appl Catal A: Gen 316 (2007) 207.

[10]-E. Kanezaki, Sol. Stat. Ion, 106 (1998) 279.

11]-L. Obalova, M. Valaskova, F. Kovanda, Z. Lacny, and K. Kolinova, Chem Pap. 58 (2004) 33.

[12]-M.R. Kang, H. M. Lim, S.C. Lee, S.H. Lee, K.J. Kim. J. Mat.Onl, 1 (2005) 1.

13]-P. Bera, M. Rajamathi, M.S. Hegde, P.V KAMATH, Bull. Mater, Sci, 2 (2000) 141.

[14]-J. Das, D. Das, K.M. Parida, J Coll. Int Sci, 301 (2006) 569.

15]-M. Marquevich, F. Medina, D. Montané, Catal Commun, 2 (2001) 119.

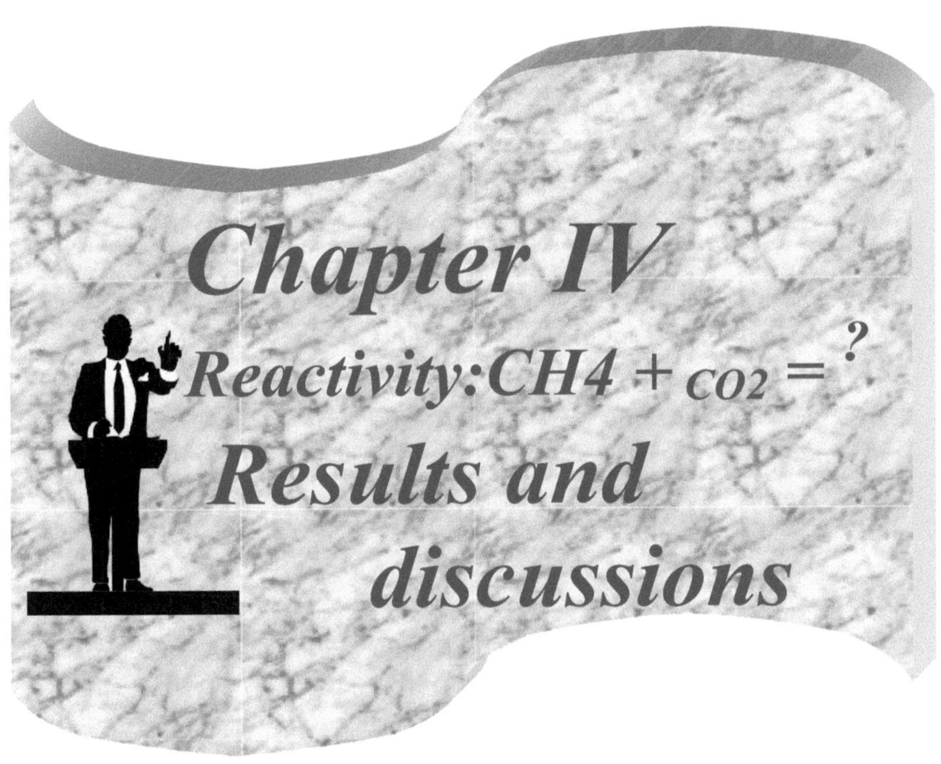

Chapter IV
Reactivity:CH4 + co2 = ?
Results and
discussions

CHAPTER IV

INTRODUCTION

The Bibliographic study in Chapter I, has shown that all Group VIII m e t a l s are suitable catalysts for methane reforming with carbon dioxide. Nevertheless, to benefit from the advantages of methane reforming (Eq.1) over water reforming, it is imperative to develop efficient and lower cost catalysts.

$$CH4 + CO_2 \longrightarrow 2CO + 2H2 \quad \textbf{(Eq.1)}$$

Some authors stipulate that catalysts derived from hydrotalcites could play an important role in the catalytic act of dry methane reforming thanks to their numerous properties, in particular their mesoporous structure and their large specific surface area.

The dry methane reforming reaction is an important means of producing synthesis gas (CO $+ _{H2}$) [1,2]. This process is all the more interesting industrially as the CO/ H2 values remain close to unity, a value required for the production of heavy hydrocarbons and oxygenated derivatives (alcohols, acids, etc.).

From an environmental point of view, the carbon dioxide methane reforming reaction (1) uses two greenhouse gases par excellence [3]. From a thermodynamic point of view, the CO_2 molecule is stable and would require reducing agents such as hydrogen or methane [4]. Hence the need for greater involvement of the dry methane reforming reaction in the various economic and industrial sectors.

The conversion of CH4 into synthesis gas is carried out in a U-shaped quartz reactor at atmospheric pressure in a differential dynamic regime, in which a catalytic charge of 0.1g is deposited.

The major problem remains the stability of the catalysts. At low temperature, the Boudouard reaction (Eq.2) is strongly favoured, leading to a more or less rapid poisoning of the catalyst.

$$\textbf{2COCO2 + C} \qquad \textbf{(Eq.2)}$$

The equilibrium of this reaction is shifted to the reagents by increasing the temperature. Unfortunately, at high temperatures, sintering of the catalysts and deactivation by cracking of the methane (Eq.3) become significant:

$$CH4 \longrightarrow C + 2H2 \quad (Eq.3)$$

Moreover, the formation of water due to the reaction of gas with water (Eq.4) accelerates this process:

$$CO2 + H2 \longrightarrow CO + H2O \quad (Eq.4)$$

The literature shows that nickel-based catalysts perform well in conventional water reforming.

The first part of this chapter is the study of the dry methane reforming reaction (Eq.1). The parameters studied are :

    -Study of the catalysts' running-in.
    -Study of the evolution of catalytic performance as a function of temperature.
    -Study of the effect of the reduction temperature for the most efficient catalysts.
    -Study of catalyst ageing and regenerability.

In addition, the catalysts with the lowest performance in dry methane reforming reactions are tested for another Friedel-Craft type reaction.

A) CARBON DIOXIDE REFORMING OF METHANE

I. STUDY OF CATALYST REGIMES

The purpose of this study is to determine the time after which the catalytic activity estimated by the overall transformation rate (OTRi) remains invariable, i.e. the catalyst's regeneration.

The catalysts are reduced overnight at T= 500°C. After the activation step, the hydrogen is replaced by the reaction mixture (CH_4 + CO_2) of composition (CH_4 , CO_2 , Ar = 20%,20%,60%).

We have thus followed the evolution of the dry reforming reaction of methane by carbon dioxide at atmospheric pressure, under the following operating conditions:

-Reaction temperature: Reaction = 650°C.

-Inlet flow rate of the reaction mixture: d $_{MR}$ = 1.5 L/ h.

-Gas ratio: CO2/ CH4 = 1.0.

-catalytic mass: m = 0.1g.

1. Study of catalysts based on MgAl-HTc hydrotalcite

The CH_4 and CO_2 conversion rates and carbon monoxide (nCO) productivity obtained in the presence of NiAl-HTc, NiMgAl-HTc and CoMgAl- HTc catalysts are shown in Figures-1-(A, B and C).

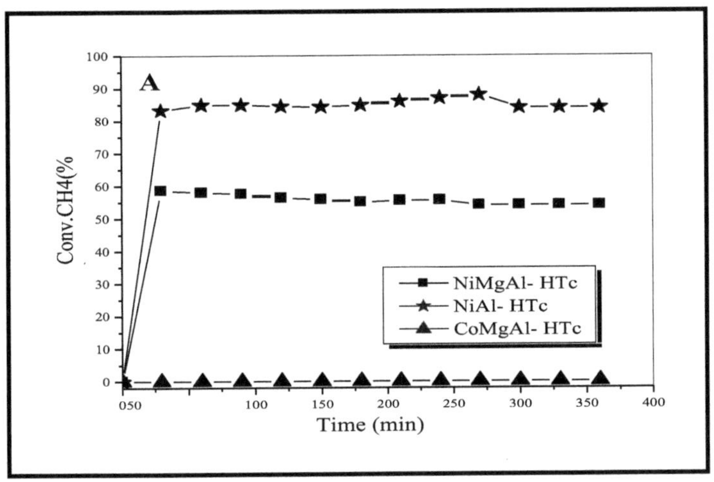
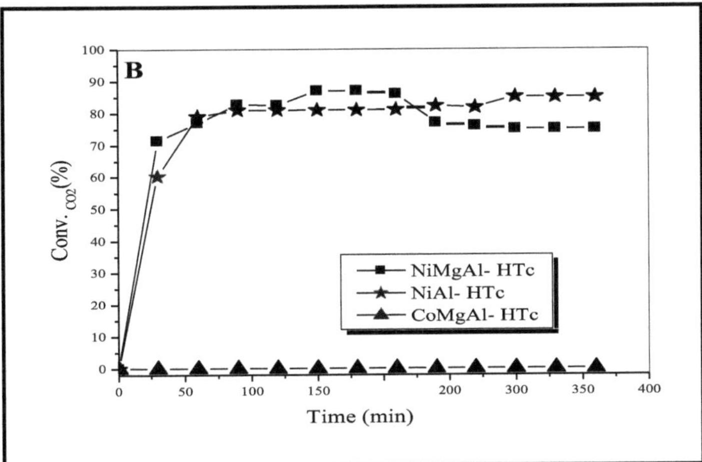

Figure-1: Evolution of the conversion of CH4 (A) and $_{CO2}$ (B) as a function of time in the presence of MgAl-HTc hydrotalcite catalysts. (Reaction = 650°C, dMR = 1.5 L/h, CO2/ CH4 =1.0).

From the previous results, it appears that catalytic activity and carbon monoxide productivity increase over time to reach a plateau representing the steady state and catalytic stability of our samples. We note that the NiMgAl-HTc and NiAl-HTc catalysts are active from the very first minutes, with CH4 conversions of 54.0% and 84.0% respectively. The CO2 conversions recorded are of the order of 75.0% and 85.0% respectively. It should also be noted that these conversions are maintained beyond 6 hours of catalyst work under reaction mixture.

However, total catalytic inertia is observed in the presence of the CoMgAl- HTc catalyst under these same working conditions.

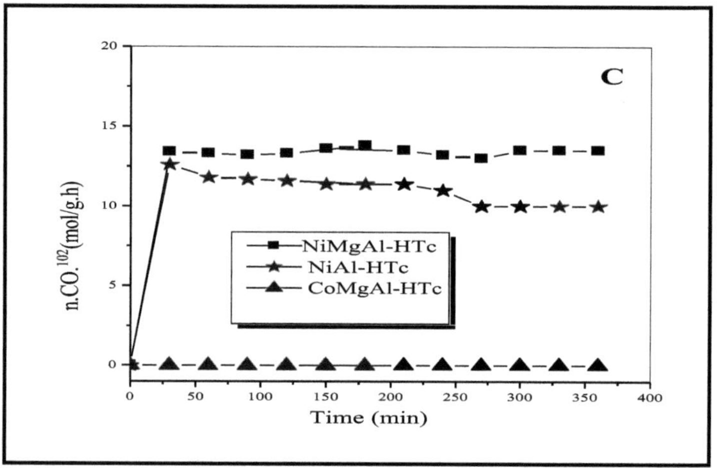

Figure-1-C: Evolution of CO productivity as a function of time, in the presence of MgAl-HTc hydrotalcite-based catalysts (Reaction = 650°C, dMR = 1.5 L/h, CO2/ CH4 =1.0).

The carbon monoxide (and hence synthesis gas) productivity, as shown in Figure-1-C, is of the order of 13.5 x 10-2 mol/ g.h. and 10.0 x 10-2 mol/ g.h. for NiMgAl-HTc and NiAl-HTc respectively.

From the results obtained, we can affirm that the stationary regime of the catalysts is established on average after 60 min of work under reaction mixture.

2. Study of MgLa series catalysts

The burn-up study of the NiMgLa-HTc, NiLa-HTc and CoMgLa-HTc catalysts is carried out under the same experimental conditions mentioned above.

The changes over time in CH4 and $_{CO2}$ conversions and carbon monoxide productivity are shown in Figures-2-(A, B and C).

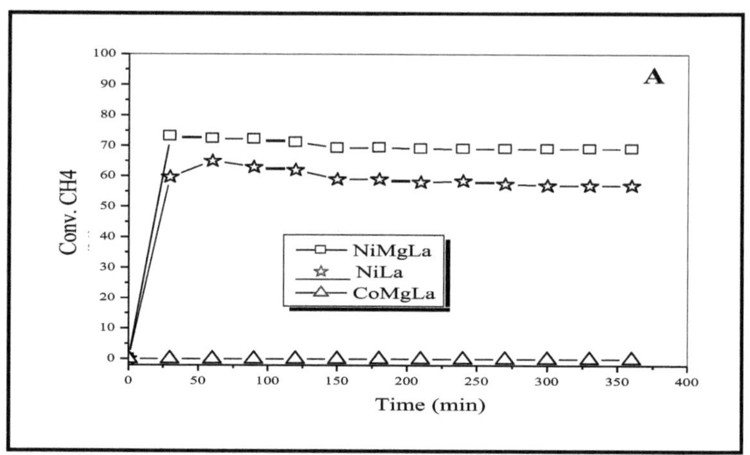

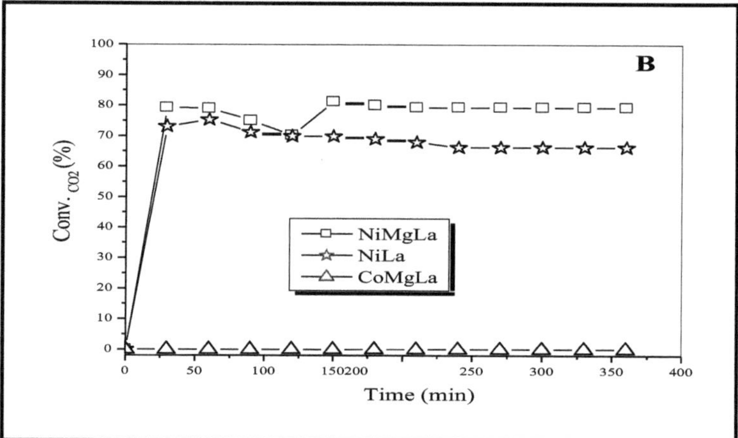

Figure-2: Evolution of CH4 (A) and CO_2 (B) conversions as a function of time, in the presence of MgLa series catalysts (Reaction = 650°C, dMR = 1.5 L/h, CO2/ CH4 = 1.0).

We note that NiMgLa and NiLa catalysts show CH4 conversions of 69.0% and 57.0% respectively, while CO_2 conversions are of the order of 79.3% and 66.3% respectively.

The carbon monoxide productivity (Figure-2-C) for these two catalysts is 11.6.10-2 mol/g.h and 11.8.10-2 mol/g.h respectively. This order of catalytic activity is maintained for more than 6 hours of work under reaction mixture.

However, the CoMgLa catalyst, as observed in the previous case, shows no catalytic activity under these same operating conditions.

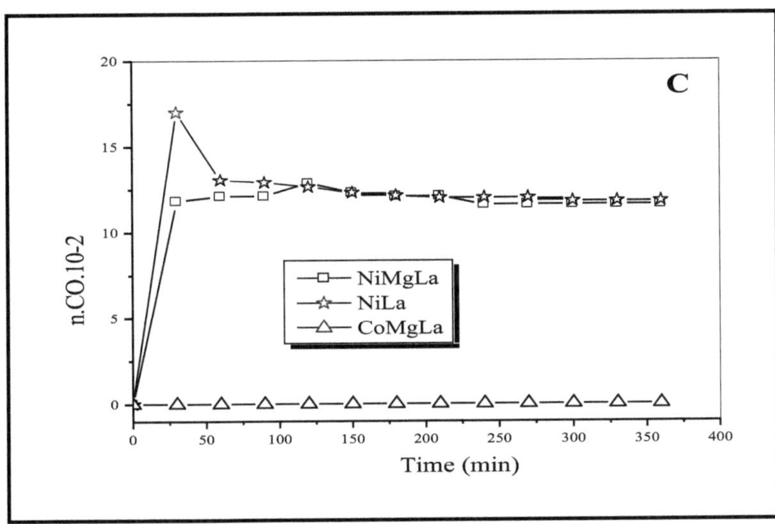

Figure-2-C: Evolution of CO productivity as a function of time, in the presence of MgLa series catalysts (Reaction = 650°C, dMR = 1.5 L/h, CO2/ CH4 =1.0).

3. Conclusion

- ➢ Under the above-mentioned operating conditions, the Ni-based samples show interesting and practically similar catalytic performances.
- ➢ CoMgAl-HTc and CoMgLa catalysts have no catalytic activity.

➤ Later in this chapter, we look for other operating conditions for reactivity of cobalt-based samples.

➤ All the samples examined are brought up to speed after 60 minutes of work under reaction mixing.

II. STUDY AS A FUNCTION OF REACTION TEMPERATURE

We examined the reaction of methane reforming with carbon dioxide in the presence of MgAl-HTc and MgLa series hydrotalcite catalysts in a temperature range of 400°C to 700°C under the following operating conditions:

-Mass of catalyst: 0.1g.
-Total flow rate of the reaction mixture: dMR = 1.5 L/h.
-Gas ratio: CO2/ CH4 =1.0 (CH4,$_{CO2}$, Ar =20%,20%,60%).
-Step of activation: Reduction during the night under hydrogen at a flow rate of 1.2 L/h at T = 500°C.

1. Study of catalysts based on MgAl-HTc hydrotalcite

The CH_4 and $_{CO2}$ conversion rates and carbon monoxide productivity obtained in the presence of NiAl-HTc and NiMgAl-HTc and CoMgAl-HTc catalysts are shown in Figures-3-(A, B and C).

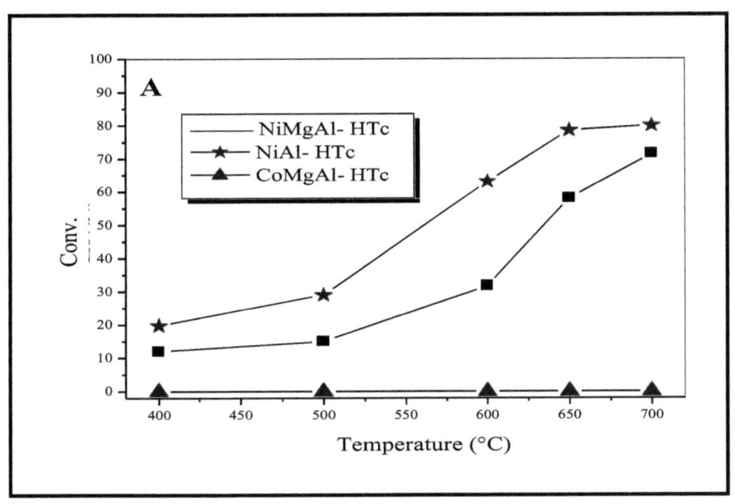

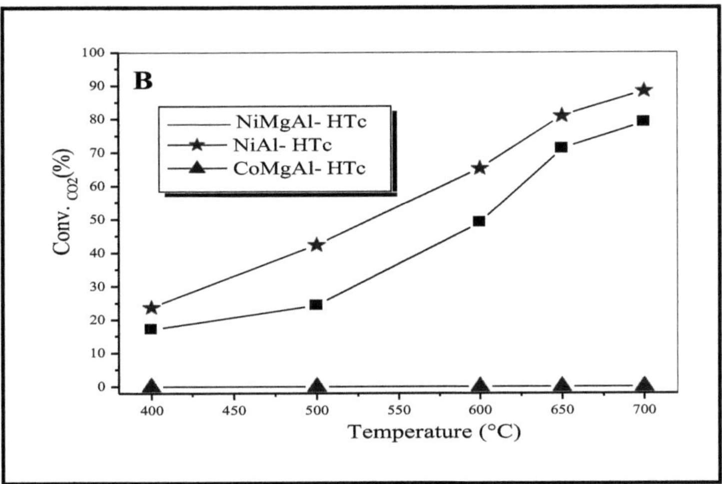

Figure-3: Evolution of CH4 (A) and $_{CO2}$ (B) conversions as a function of reaction temperature in the presence of MgAl-HTc hydrotalcite catalysts. (dMR = 1.5 L/h, CO2/ CH4 =1.0).

We note firstly that the CoMgAl-HTc catalyst does not show any catalytic activity under the operating conditions mentioned above, whatever the reaction temperature ranging from 400°C to 700°C.

On the other hand, NiMgAl-HTc and NiAl-HTc catalysts show a significant catalytic activity at a relatively low temperature (400°C). CH4 conversions are then of the order of 12.0% and 19.8% respectively.

At 650°C, conversions to synthesis gas are significantly improved, i.e. the conversion of CH4 in the presence of NiMgAl-HTc catalysts, NiAl-HTc reaches 58.1% and 78.6% respectively.

The carbon monoxide productivity nCO (Figure-3-C) increases from 1.35 x 10-2 mol/g.h to 15.0 x 10-2 mol/g.h when the temperature increases from 400°C to 650°C in the case of the NiMgAl-HTc catalyst and from 4.1 x 10-2 mol/g.h to 13.1 x 10-2 mol/g.h in the case of the NiAl-HTc solid, thus confirming the highly endothermic nature of the reaction [5].

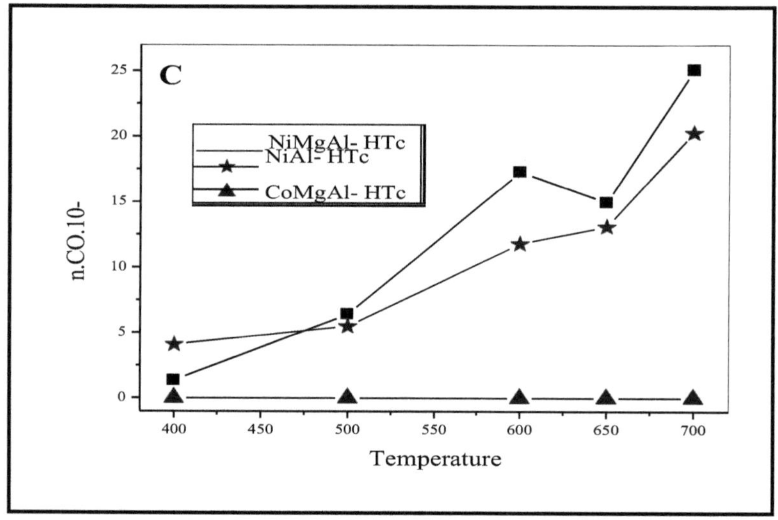

Figure-3-C: Evolution of CO productivity as a function of reaction temperature with MgAl-HTc hydrotalcite catalysts (dMR = 1.5 L/h, CO2/CH4 =1,0).

Furthermore, the two reagents CH4 and CO_2 seem to facilitate each other's dissociation as Erdöhely et al [6] found.

We can also note that the conversions of the two reagents are of the same order of magnitude in the case of the solid NiAl-HTc, which would suggest the absence of side reactions at 650°C.

On the other hand, at higher temperatures (700°C), CO_2 conversions are significantly higher than CH4 conversions in the case of NiMgAl-HTc and NiAl-HTc samples.

Namely, CH4 and CO_2 conversions are 71.5% versus 79.0% and 80.0% versus 88.4% respectively in the presence of the NiMgAl-HTc and NiAl- HTc catalyst. These results suggest that secondary reactions are taking place, in particular, the gas-to-water reverse reaction [7]:

$$CO_2 + H_2 \longrightarrow CO + H_2O$$

Note that significant amounts of water are collected in the water trap during this study.

The carbon monoxide productivity evolves in the same direction, i.e. when the temperature increases from 650°C to 700°C, the CO productivity increases from 15.0.10-2 mol/g.h to 25.1.10-2 mol/g.h, and from 13.1.10-2 mol/g.h to 20.3.10-2 mol/g.h, respectively in the presence of NiMgAl-HTc and NiAl-HTc catalysts.

As for the theoretical balance of carbonaceous products, it goes from 81.0% to 82.0% in the case of NiMgAl-HTc, and from 85.3% to 82.5% in the case of NiAl-HTc when the temperature goes from 650°C to 700°C suggesting a slight carbon deposit whose origin could be [8,9] :

➢ CH4 decomposition (Eq.3) :

$$CH4 \longrightarrow C + 2 H_2$$

➢ Or to the CO dismutation reaction (Eq.4) :

$$2 CO \longrightarrow C + CO_2$$

On the other hand, we note that the NiMgAl-HTc catalyst has superior catalytic performance compared to solid NiAl-HTc. This phenomenon could be due to the presence of Mg^{2+} cations in the framework of the solid.

The presence of magnesium in the catalyst structure, as shown in the literature, decreases the sensitivity of the catalyst to coke formation [10]. Some authors even assume that the formation of a NiO-MgO solid solution after calcination could improve the catalytic performance of the sample [11].

In this context, Battacharya et al [12], compared oxide catalysts from hydrotalcites with conventional Ni/Al2O3 and Ni/MgAl2O4 catalysts. They showed that hydrotalcite-derived oxides had better catalytic performance in terms of reactivity and resistance to carbon deposition in dry methane reforming reactions.

These results also show that a better dispersion of the active phase is ensured when hydrotalcites are used as bases or catalytic supports.

2. Study of MgLa series catalysts

Figures-4-(A, B and C) show the evolution of methane, carbon dioxide and carbon monoxide productivity as a function of temperature in the presence of MgLa series catalysts. The catalytic test is carried out under the same operating conditions already mentioned.

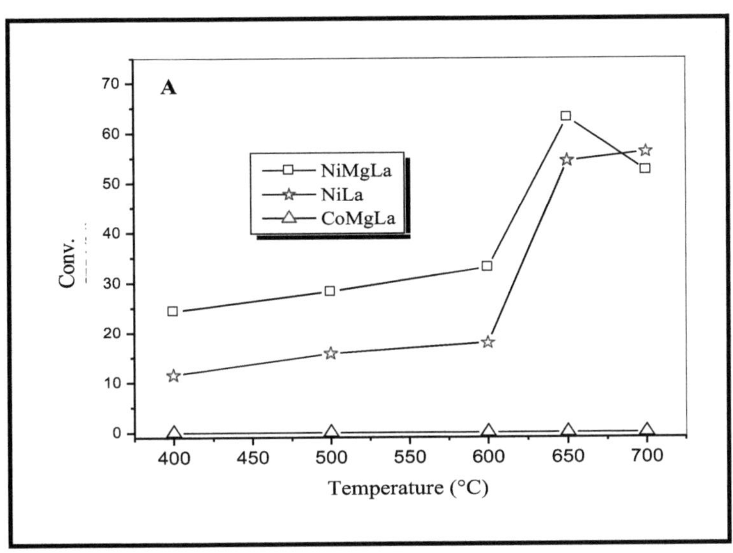

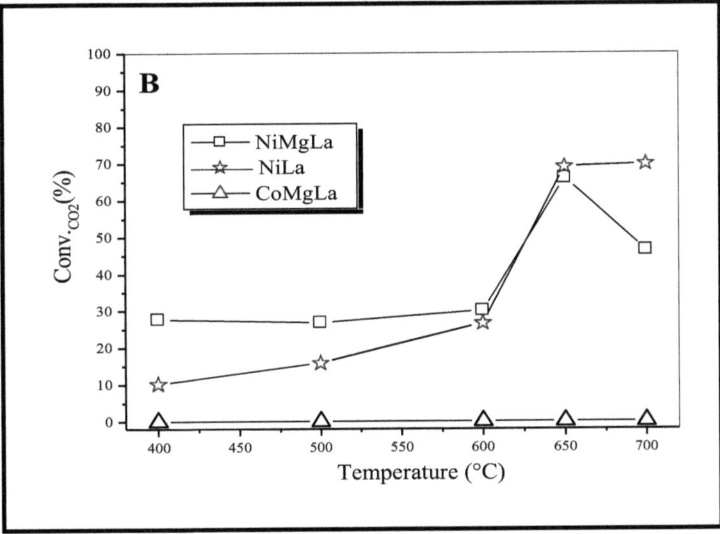

Figure-4: Evolution of CH4 (A) and CO_2 (B) conversions as a function of reaction temperature in the presence of MgLa series catalysts (dMR = 1.5 L/h, CO_2/CH_4 = 1.0).

As in the previous case, we have noted that the CoMgLa catalyst has total catalytic inertia regardless of the reaction temperature ranging from 400°C to 700°C.

In addition, NiMgLa and NiLa catalysts show a significant CH_4 and CO_2 conversion at 400°C, i.e. 24.3% compared to 27.7% and 11.5% compared to 10.2% respectively.

At 650°C, CH_4 and CO_2 conversions reached 63.0% and 54.4%, respectively, compared to 66.0% for NiMgLa-HTc and 69.1% for NiLa-HTc.

Similarly, carbon monoxide productivity (Figure-4-C) increases from 2.7.10-2 mol/g.h to 10.8.10-2 mol/g.h in the case of the NiMgLa sample, and from 1.5.10-2 mol/g.h to $16.5.^{10-2}$ mol/g.h in the case of NiLa when the temperature increases from 400°C to 650°C.

Moreover, both catalysts have higher CO_2 conversions than CH_4, suggesting that secondary reactions are taking place, in particular the reverse reaction of gas with water (RWGS).

The theoretical carbonaceous product balance increases from 83.6% to 74.0% when the temperature rises from 400°C to 650°C and from 92.1% to 70.0% in the case of NiMgAl-HTc and NiLa catalysts respectively.

However, the study of the nature and origin of carbon would require further and more elaborate studies.

Furthermore, a comparison of all NiMgAl-HTc, NiAl-HTc, NiMgLa and NiLa nickel-based catalysts at 650°C leads us to conclude that no significant differences are observed.

At higher temperature (700°C), the CO_2 and CH_4 conversions recorded in the presence of the different catalysts are respectively 79.0% versus 71.5% in the case of NiMgAl- HTc, 88.4% versus 80.0% in the presence of NiAl-HTc, and 69.8% versus 56.2% in the case of NiLa catalyst and 46.4% versus 52.4% in the case of NiMgLa.

Similarly, the carbon monoxide productivity at 700°C is 25.1.10-2 mol/g.h, 20.3.10-2 mol/g.h, 14.2.10-2 mol/g.h and 19.6.10-2 mol/g.h respectively in the presence of NiMgAl-HTc, NiAl-HTc, NiLa and NiMgLa catalysts.

These results clearly show that the NiMgAl-HTc catalyst has better catalytic performance compared to its counterparts.

Thus, at 700°C, we can classify all of these solids according to their decreasing catalytic performance according to the following sequence: NiMgAl-HTc > NiAl- HTc > NiMgLa > NiLa.

On the other hand, the basic character of our samples would evolve in a decreasing direction as shown in the following sequence: NiMgLa > NiLa > NiMgAl-HTc > NiAl-HTc, since La3+ cations [have] a higher basic character than Mg2+ ions and Al3+ ions have an acidic character·

It can be argued that better catalytic performance is expected when samples have an intermediate acid-base nature as is the case with the NiMgAl-HTc sample [13].

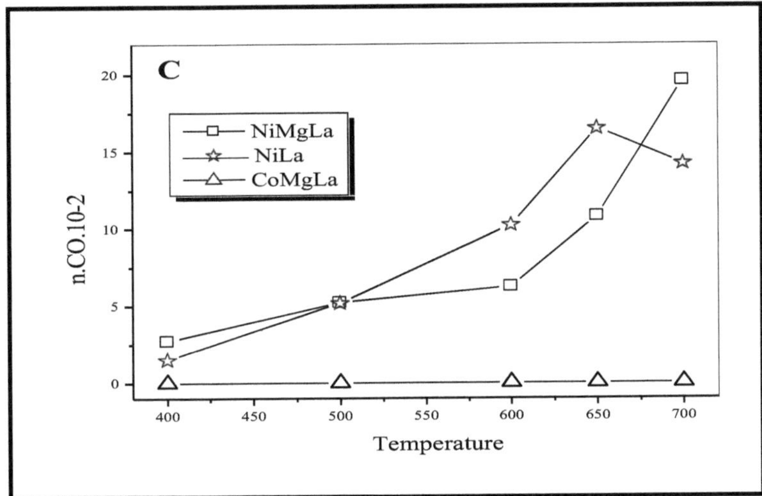

Figure-4-C: Evolution of CO productivity as a function of the reaction temperature in the presence of MgLa series catalysts (dMR = 1.5 L/h, CO2/CH4 =1,0).

It is also necessary to point out that all the CrMgAl-HTc, CuMgAl-HTc and FeMgAl-HTc samples do not show any catalytic activity in dry methane reforming reaction, which led us to test them for another Friedel-Craft type catalytic reaction.

3. Conclusion

➢ In this study, we showed essentially the effect of the nature of the transition metal, i.e., under our operating conditions, only nickel-based catalysts are active, so that the CoMgAl-HTc and CoMgLa-HTc catalysts remain totally inactive whatever the reaction temperature.

➢ MgAl-HTc hydrotalcite-based catalysts have better catalytic performance than those based on the MgLa series.

➢ All the nickel-based samples have catalytic performances at 700°C, which evolve according to the following sequence: NiMgAl-HTc > NiAl-HTc > NiMgLa > NiLa.

III. STUDY OF THE EFFECT OF REDUCTION TEMPERATURE

Since the reducibility of nickel ions is a key factor in reactivity, synthesis gas productivity and stability, we felt it was necessary to study this parameter and its influence on the reactivity of our samples. We also considered it useful to modify this factor in order to re-examine the reactivity conditions of Co-based catalysts.

The experimental conditions are:

> *-Mass of catalyst: 0.1g.*
> *-Total flow rate of the reaction mixture: dMR = 1.5 L/h.*
> *- Gas ratio: CO2/ CH4 =1 0. (CH4, $_{CO2}$, Ar = 20%,20%,60%).*
> *-Step of activation: Reduction during the night under hydrogen at*
> *a flow rate of 1.2 L/h at T = 500°C.*

The best performing NiMgAl-HTc, NiAl-HTc and CoMgAl-HTc catalysts will thus be examined in different situations: without reduction, after reduction at 500°C and 650°C (θ = 4°C/min) under hydrogen (1.2 L/ h)

1. Study of the NiMgAl-HTc catalyst

The CH_4 and CO_2 conversion rates and CO productivity obtained for the NiMgAl-HTc catalyst are shown in Figures-5-(A, B and C).

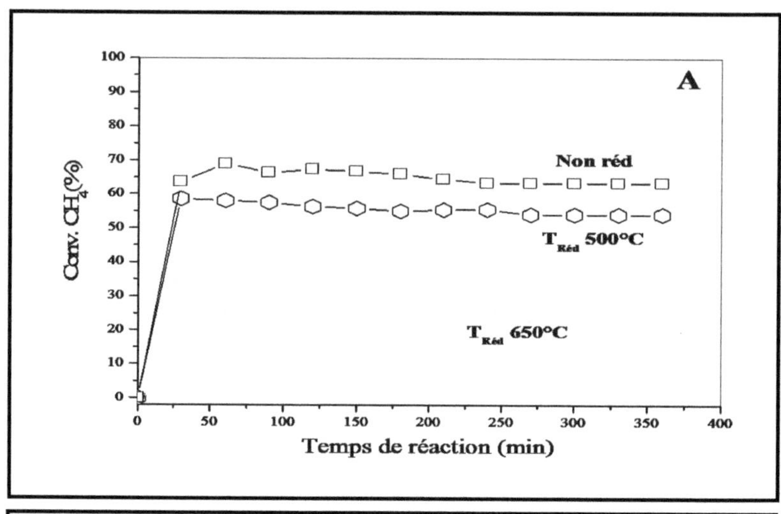

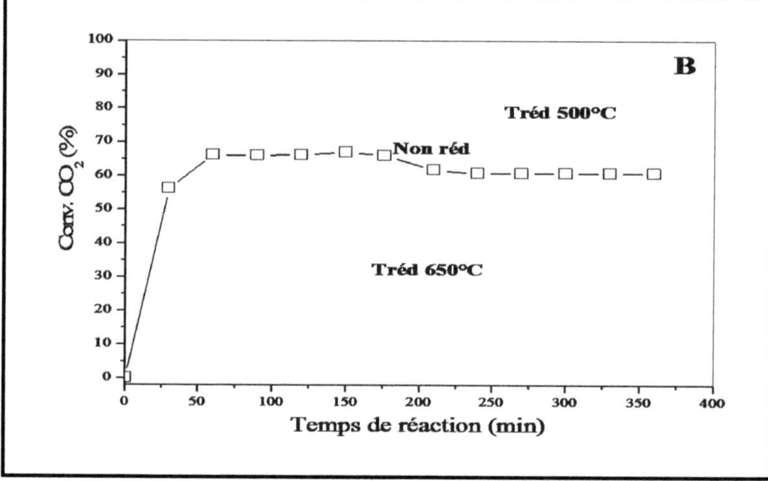

Figure-5: Conversions of CH4 (A) and $_{CO2}$ (B), in the presence of NiMgAl- HTc catalyst at different reduction temperatures (Reaction = 650°C, dMR = 1.5 L/h, $_{CO2/CH4}$ = 1).

We note , essentially, that the catalyst NiMgAl-HTcpresents significantly better catalytic performance when not reduced.

The CH4 conversions are then of the order of 63.3% when the sample is not reduced and 54% and 25% when it is reduced at 500°C and 650°C respectively.

CO_2 conversions appear slightly higher when the sample is reduced to 500°C. They are about 74.0% and 27.0% when the sample is reduced to 500°C and 650°C respectively. Whereas it is of the order of 60.7% when the sample is not reduced.
It can also be noted that the carbon monoxide productivity is higher in the case of the non-reduced catalyst (24.8.10-2 mol/g.h). It is of the order of 13.2.10-2 mol/g.h and 5.10-2 mol/g.h respectively when the solid NiMgAl-HTc is reduced at 500°C and 650°C.
The low catalytic activities recorded in the case of the sample reduced to 500°C could be attributed to the formation of spinel species that are difficult to reduce under our operating conditions [12].

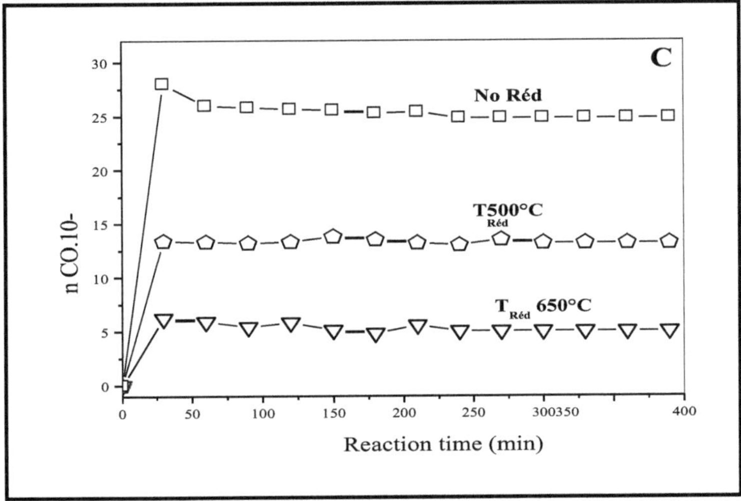

Figure-5-C: CO productivity in the presence of reduced NiMgAl-HTc catalyst at different reduction temperatures (Reaction = 650°C, dMR = 1.5L/h, CO_2/CH_4 =1).

2. Study of the NiAl-HTc catalyst

The CH4 and CO_2 conversion rates and carbon monoxide (CO) productivity obtained from the NiAl-HTc catalyst are shown in Figures-6-(A, B and C).

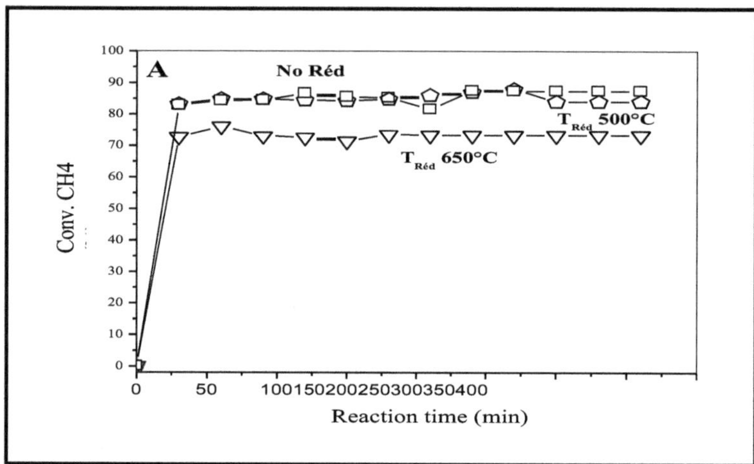

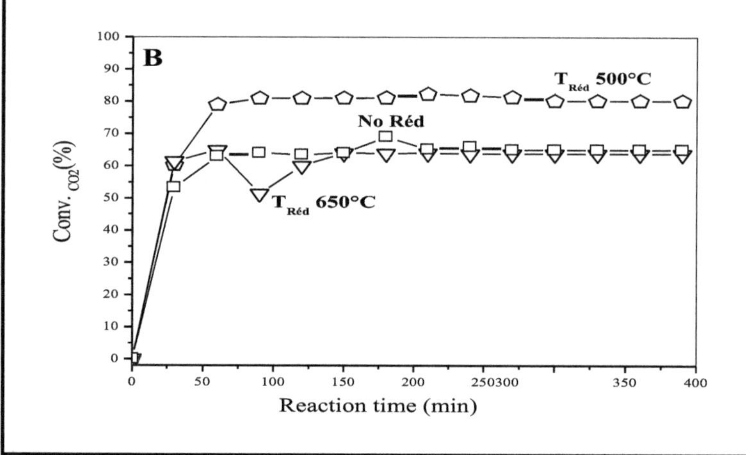

Figure-6: Conversion of CH4 (A) and CO_2 (B) in the presence of the reduced NiAl-HTc catalyst at different reduction temperatures (Reaction= 650°C, dMR = 1.5L/h, $_{CO2/CH4}$ =1).

The NiAl-HTc catalyst has a different catalytic pathway than its NiMgAl-HTc counterpart. Thus, the CH4 conversions obtained in the case of the non-reduced catalyst. They are of the order of 87.4% in the case of the non-reduced catalyst. reduced at 650°C, are of the order of 73.3%.

On the other hand, CO_2 conversions appear to be higher than CH4 conversions, in the case of the sample reduced to 500°C. It reaches a value of the order of 80.1%.

Carbon monoxide productivity follows the same trend. It is around 21.5. 10-2 mole/g.h in the case of the unreduced sample as against an average value of 10,0.10-2 mole/g.h in the case of the catalyst reduced to 500 ^{0C} and 650 ^{0C}.

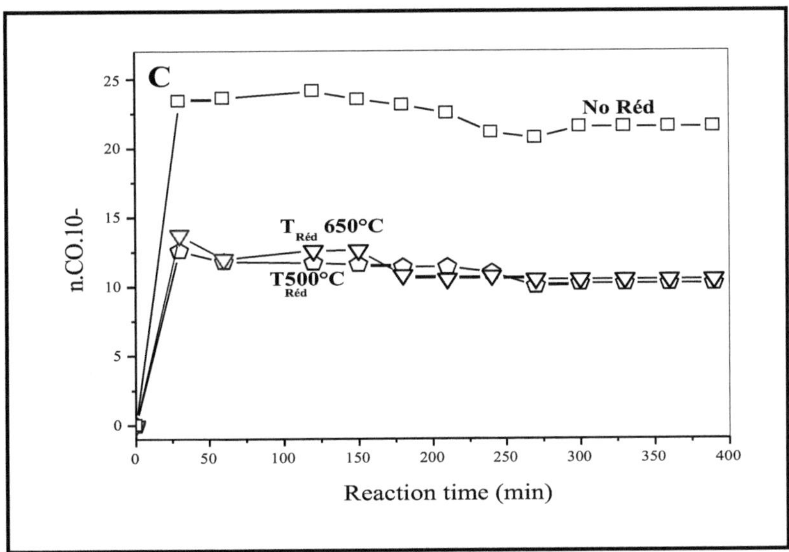

Figure-6-C: CO productivity in the presence of reduced NiAl-HTc catalyst at different reduction temperatures. (Reaction = 650°C, dMR = 1.5l/h, CO_2/CH_4 =1.0).

As we have seen, unreduced NiMgAl-HTc and NiAl-HTc solids exhibit significant catalytic activity. After their reduction, the results show that the conversion of methane is significantly less than that of carbon dioxide. This can be explained either by the presence of the secondary reaction, in particular the reverse reaction of gas with water (RWGS).

3. Study of the CoMgAl-HTc catalyst

The same study as above was carried out for the CoMgAl-HTc catalyst. This study will examine both the effect of the reduction temperature and the effect of the nature of the M2+ cation (Ni2+ and Co2+) on the catalytic performance of these samples. The results are shown in Figures-7-(A, B and C).

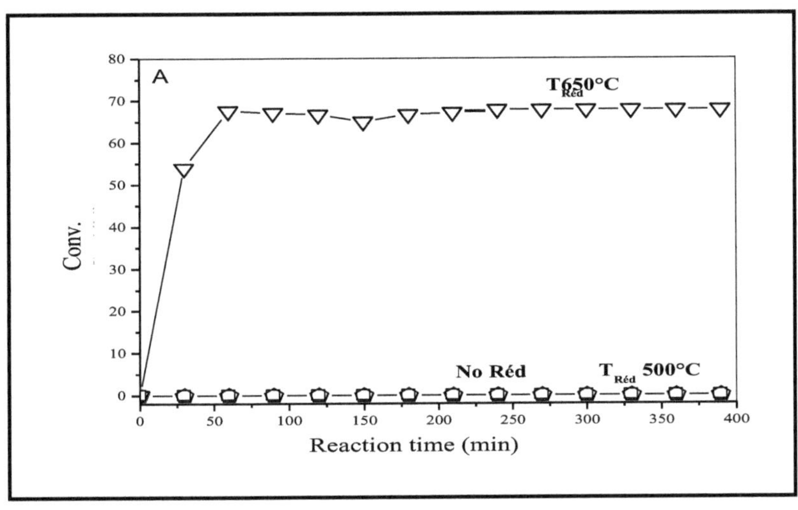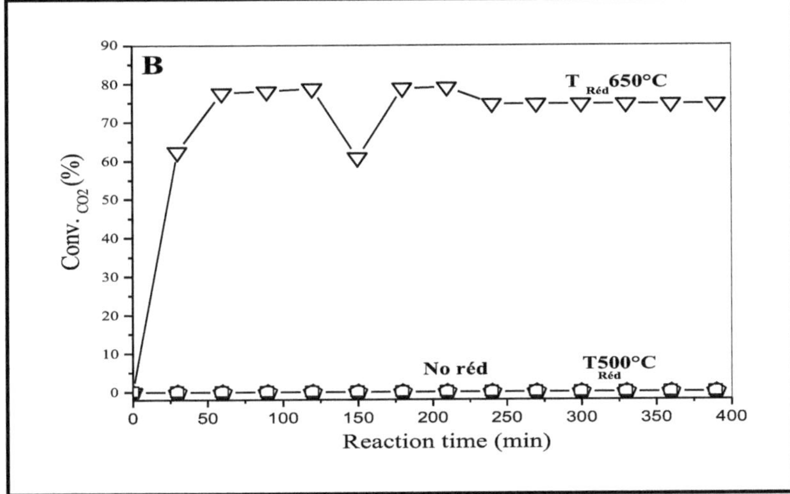

Figure-7: Conversion of CH4 (A) and $_{CO2}$ (B) in the presence of the reduced CoMgAl-HTc catalyst at different reduction temperatures. (Reaction= 650°C, dMR = 1.5L/h,$_{CO2/CH4}$ =1)

The CoMgAl-HTc catalyst exhibits total catalytic inertia when not reduced or when reduced to 500 ^{0C} (already demonstrated in the temperature effect study).

On the other hand, it shows significant CH4 and $_{CO2}$ conversions when reduced to 650°C. In the latter case, the CH4 and $_{CO2\ conversions}$ are of the order of 67.8% and 74.7% respectively. The carbon monoxide productivity is therefore of the order of 15.0.10-2 mol/g.h. The CO/H2 ratio shows values close to unity, which allows us to confirm that secondary reactions do not occur under these experimental conditions.

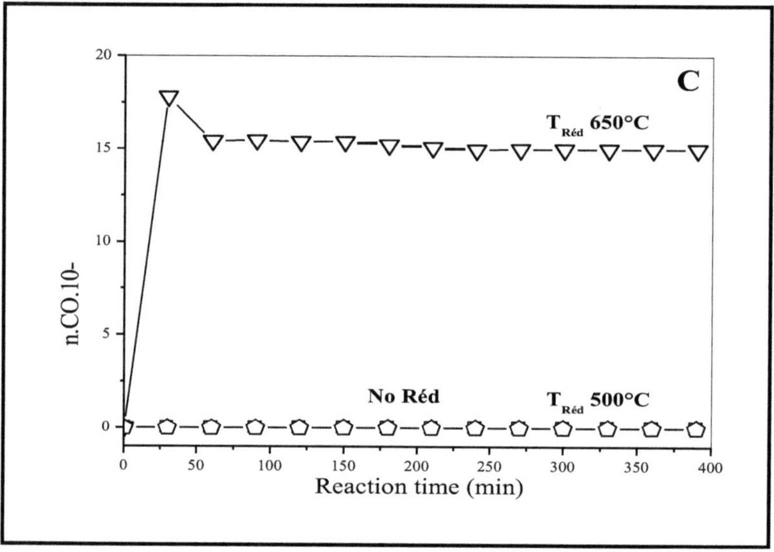

Figure-7-C: CO productivity for the CoMgAl-HTc catalyst over time, reduced at different temperatures (Reaction = 650°C, dMR = 1.5l/h, $_{CO2/CH4}$ =1).

4. Conclusion

- This study essentially examined the optimal reactivity conditions for the cobalt sample.
- Thus, we have shown that reduced to 650°C, significant CH4 and CO_2 conversions are recorded. On the other hand, when it is not reduced or when it is reduced at 500°C, no catalytic activity is observed.
- It has also been shown that NiAl-HTc and NiMgAl-HTc catalysts show better performance when not reduced under H_2 and that unreduced catalysts have superior catalytic performance.

IV. AGEING AND REGENERATION OF CATALYSTS

INTRODUCTION

Among the most important phenomena that can cause catalyst deactivation are sintering of the active phase, carbon deposition and sulphur poisoning. We will focus our study on the first two phenomena only, as they are closely linked to operating conditions, whereas deactivation by sulphur is more a matter of purifying the feedstock by desulphurisation units, a solution that is now available in all petrochemical units.

1. Sintering

Sintering of the active phase results in changes in structure and morphology under the influence of several parameters such as time, temperature and reaction atmosphere. Thus the initially well dispersed metal particles increase in size and the catalyst loses its activity. In general, sintering phenomena are observed in very high temperature ranges.

2. Carbon Deposition

The formation of carbonaceous deposits from the reagents can lead to surface fouling, catalyst pore blockage or carrier disintegration.

Many authors, including Reitmeier et al [14], White et al [15], Sacco et al [16], have presented thermodynamic calculations to predict the formation of carbon deposition as a function of operating conditions. The main conclusion of this study is the need to work at temperatures above 727°C with a CO_2/CH_4 ratio much higher than unity in order to avoid the thermodynamic zone of carbon formation. However, for the application of this process in the industrial field, it is recommended to work at a lower temperature with a CO_2 / CH4 ratio close to the unit.

Pailleurs, it is mentioned in many works, that the origin of the inactive carbon formed during the dry reforming of methane, is either the decomposition reaction of methane which is endothermic, or the dismutation reaction of CO which is exothermic.

The carbon formed during dry methane reforming is often in filamentary form. Rodiguez stated in a review that the critical phase in the formation of this carbon is the diffusion of carbon into metal particles [17].

In addition, the formation of carbonaceous deposits includes the production and transformation of several forms of carbon [18].

Koerts et al [19], Sacco et al [16], Jablonski et al [20] and Zhang et al [21] identify three types of carbon on supported transition metals: Cα, Cβ, Cγ. Among these types of carbon, the active species Cα would be responsible for the formation of CO while the less active carbons Cβ and Cγ would be responsible for the deactivation of the catalyst. Koerts et al [19] state that Cα also called carbidic carbon is easily removed by hydrogenation and that Cβ or amorphous carbon, is reduced at a temperature close to 300°C, while Cγ, also called graphitic carbon, is removed at a temperature higher than 400°C.

3. Study of catalyst ageing

In the preceding context, we have studied the ageing of catalysts according to the cycle described in the diagram below where a1, a2, a3, and a4 represent the different CH4 and CO_2 conversions obtained at the different stages of the thermal cycle, in the presence of the NiMgAl-HTc, NiAl-HTc and CoMgAl-HTc catalysts. Table-1 and Table-2 combine the results obtained in this study.

The operating conditions:

-Mass of catalyst = 0.1g.

-DH2 = 1.2 L/h.

-Reduction temperature = 650°C (θ = 4°C/min)

-Reduction time = 12 h.

-Gas ratio= CH4 / $_{CO2}$ =1.

After 12 hours of activation under hydrogen, the hydrogen is replaced by the reaction mixture ($_{CO2}$, CH4). A conversion measurement is carried out at 500°C (**a1**). Subsequently, the catalyst undergoes a rise in reaction temperature (4°C/min) up to 700°C, the catalytic activities are recorded (**a2**). The catalyst is then aged for 12 hours under a reaction mixture. An activity measurement is carried out at the end of this stage (**a3**). An activity (**a4**) is recorded at 500°C, after the ageing phase.

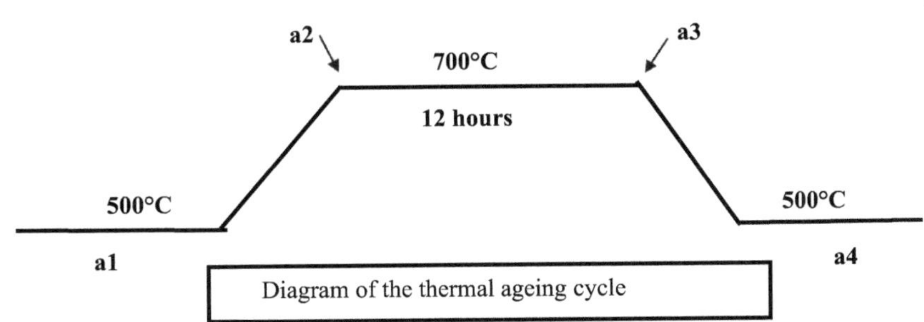

Diagram of the thermal ageing cycle

Table-1: Evolution of CH4 and $_{CO2}$ conversions during the ageing cycle.

Conversions (%)	a1		A2		a3		a4	
	CH4	CO2	CH4	CO2	CH4	CO2	CH4	CO2
NiMgAl-HTc	23,2	21,8	61,0	62,2	55,0	85,2	00,0	00,0
CoMgAl-HTc	00,0	00,0	23,3	23,8	25,7	31,5	00,0	00,0
NiAl-HTc	14,2	18,3	87,0	85,0	93,7	94,4	58,0	62,4

Table-2: Evolution of CO productivity during the ageing cycle.

Catalyst	Productivity in CO			
	a1	a2	a3	a4
NiMgAl-HTc	0,85	12,20	12,19	0
CoMgAl-HTc	0	2,1	1,55	0
NiAl-HTc	2,10	12,74	18,02	5 ,11

The CH4 conversions recorded at 500°C (a1), in the presence of NiMgAl- HTc, NiAl-HTc and CoMgAl-HTc catalysts are of the order of 23.2%, 14.2% and 00.0% respectively.

The carbonaceous product balances are practically close to 100% at this same temperature. At 700°C, before the ageing phase (**a2),** the CH4 and $_{CO2}$ conversions reached 61.0% against 62.2%; 23.3% against 23.8% and 87.0% against 85.0% respectively in the presence of the catalysts NiMgAl-HTc, CoMgAl-HTc and NiAl-HTc.

After aging at 700°C overnight, CH4 and $_{CO2}$ conversions do not seem to decrease. They reach 55.0% against 85.2%, 25.7% against 31.5% and 93.7% against 94.4% in the presence of NiMgAl-HTc, CoMgAl- HTc and NiAl-HTc catalysts respectively. As we can see, the NiMgAl-HTc catalyst seems to favour the development of secondary reactions, in particular, the reverse reaction of gas with water (Conv. $_{CO2}$ > Conv.CH4).

At 700°C before ageing, carbon monoxide (nCO) productivity increases to 12.2 x 10-2 mol/g.h, 12.7 x 10-2 mol/g.h and 2.1 x 10-2 mol/g.h in the presence of NiMgAl-HTc, NiAl-HTc and CoMgAl-HTc catalysts respectively.

Calculation of the theoretical carbon balance for the NiMgAl-HTc catalyst shows a remarkable carbon deposit of 36.0%. However, the carbon monoxide productivity does not seem to decrease and reaches on the contrary 12.2.10-2 mol/g.h.

This result could suggest that the carbon formed during the ageing phase would be of a non-toxic nature (such as Cα), as reported in the literature [22].

We estimate the degree of ageing, thanks to the Ri report. With :

Ri = [Conversion after aging (a4) / Conversion before aging (a1)].

Table-3: Evaluation of the ageing rate at 700°C.

Catalysts	NiMgAl-HTc	NiAl-HTc	CoMgAl-HTc
CHR4	00	4,1	00
BCR2	00	3,4	00
R n $_{CO}$	00	2,43	00

Ageing is all the more important as the Ri ratio is less than unity (Ri < 1). After the ageing phase and return to 500°C, the NiAl-HTc catalyst is the only sample that resists ageing and whose CH4 and $_{CO2}$ conversions are of the order of 58.0% and 62.2% respectively. As a result, it shows CH4 and CO2 conversions as high as before the ageing phase, whereas its NiMgAl-HTc and CoMgAl-HTc counterparts no longer show any catalytic activity after this drastic ageing cycle.

X-ray diffraction analysis of the NiAl-HTc sample after the aging cycle (Figure-8) showed primarily the formation of NiO species.

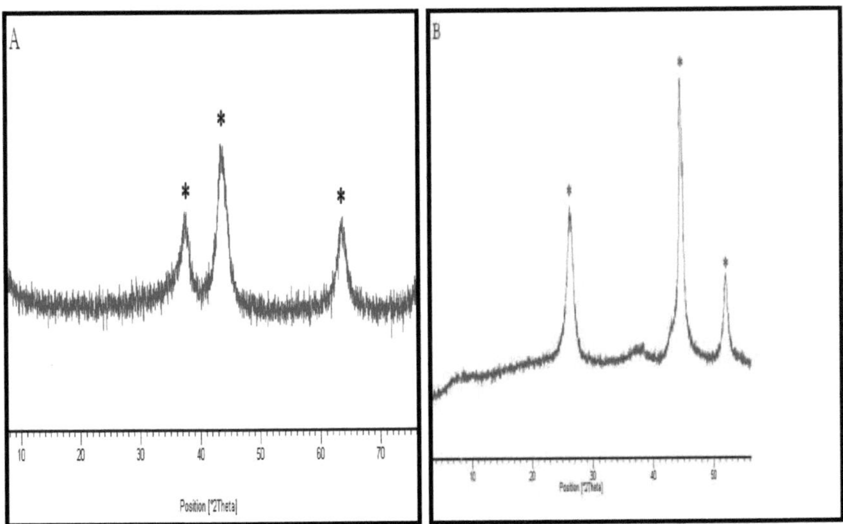

Figure-8: X-ray Difractogram of the NiAl-HTc sample. Fresh (A) and used (B).

* NiO phase

Similarly, carbon monoxide productivity decreases to 5.1.10-2 mol/g.h.

The deactivation of the catalysts under our operating conditions could most probably be due to carbon deposition and/or chemical sintering of the active phase.

Analysis of the sample by electron microscopy reveals carbon deposition (in filamentary form) as shown in Figure-9. The origin of this carbon would be either :

Boudouard reaction [23] : $2CO \longrightarrow Cs + CO_2$.

Or the decomposition reaction of CH4 [24]: $CH4 \longrightarrow C + 2H2$.

Figure-9: SEM image of the NiAl-HTc catalyst after the ageing phase

However, no hypothesis could be put forward at this level of our work. Further investigations and studies are necessary for a better understanding of the catalytic action of our samples, in particular the solid NiAl-HTc.

B)STUDY OF THE ALKYLATION REACTION OF BENZENE WITH BENZYL CHLORIDE

INTRODUCTION

Friedel-Craft reactions in particular the alkylation reactions of aromatic compounds with aromatic halides lead to polyaromatic compounds which are considered to be the basis of classical organic chemistry. These reactions are usually used in homogeneous catalysis, using Lewis acids in the liquid phase such as $FeCl_3$, or transition metal or rare earth chlorides in stoichiometric amounts [25]. Indeed, the application of these catalysts in homogeneous catalysis poses some problems. Among other things, they are irrecoverable because they very often form complexes with the alkylation reagents or with the products.

To overcome these difficulties, it seemed advantageous to examine these reactions in the field of heterogeneous catalysis, using solid acids such as: zeolites (HY) [26], clays (montmorillonite K10) [27], mesoporous silicate materials such as MCM-41 and HMS [28,29], which do not form stable complexes with the products, which allows at the end of the reaction to easily regenerate and reuse the catalyst, and to reduce the corrosion of the installations.

However, the use of basic solids in the alkylation reaction of benzene with benzyl chloride remains rare.

In this context, we have tried to determine the catalytic properties of our basic solids, namely: activity, stability and selectivity in this reaction, which is carried out according to the following scheme:

wherein n1 and n2 are 0.1 and (n1 + n2) is less than or equal to 3.

Benzene + Benzyl Chloride   ⟶   ♠ Diphenyl methane (n1+ n2 = 0).

                                       ♠ Dibenzyl benzene (n1+ n2= 1).

                                       ♠ Tribenzyl benzene (n1+ n2= 2).

I. RESULTS AND DISCUSSIONS

In this part, we first studied the alkylation reaction of benzene with benzyl chloride at Trea = 80°C in the presence of the following calcined catalysts : NiMgAl-HTc, CoMgAl-HTc, CuMgAl-HTc, CrMgAl-HTc and FeMgAl-HTc (mcat = 0.1g) with a molar ratio Bz/ClBz = 15.0.

The activation of our solids consists in pre-treating them before reaction, under air for 3 hours at a temperature of 300°C. The pre-treatment temperature is reached after 60 min of heating. The results of this study are shown in Figure 10 and in Table 3.

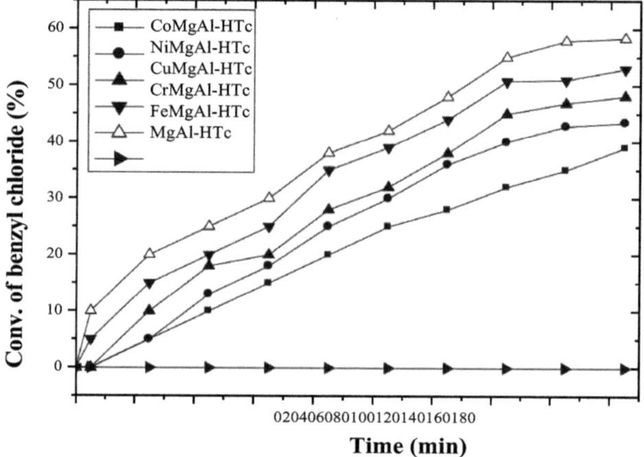

Figure-10: Evolution of the conversion of benzyl chloride as a function of the reaction time at Trea = 80°C of the MeMgAl-HTc series with Me = Co, Ni, Cu, Cr, Fe.

In the light of the results, we find that the basic solid MgAl-HTc is completely inactive in the alkylation reaction of benzene with benzyl chloride.

However, the incorporation of transition elements such as iron, nickel, chromium, copper and cobalt in the framework of MgAl-HTc hydrotalcite remarkably increases the catalytic activity of this solid.

Table-3: The catalytic properties of calcined solids type MeMgAl-HTc with Me = Co, Ni, Cu, Cr, Fe (The ratio of benzene to benzyl chloride = 15. Reaction temperature = 80°C, reaction time = 3h).

Catalysts	Conversion of chloride of benzyl (%)	Selectivity (%) [a]	
		Diphenyl methane	Poly benzyl Benzene
MgAl-HTc	-	-	-
CoMgAl-HTc	39,0	100	-
NiMgAl-HTc	43,4	100	-
CuMgAl-HTc	48,1	100	-
CrMgAl-HTc	51,0	99,8	0,2
FeMgAl-HTc	58,3	98,9	1,1

[a] Selectivity recorded at 3 hrs reaction time.

Among the samples tested, the solid FeMgAl-HTc is the one with the highest catalytic activity. Indeed, it shows a high conversion of benzyl chloride (58.3%) as well as a good selectivity to diphenyl methane (98.9%).

The catalytic activity of the samples studied follows the following ascending order:

CoMgAl-HTc < NiMgAl-HTC < CuMgAl-HTC < CrMgAl-HTC < FeMgAl-HTc.

A very high selectivity in diphenyl methane is obtained in the presence of all catalysts (100%). Nevertheless, traces of by-products are observed for both CrMgAl-HTC and FeMgAl-HTC catalysts.

It should be noted that the CuMgAl-HTc, CrMgAl-HTc and FeMgAl-HTc catalysts, which are totally inactive in dry methane reforming reactions, have interesting performances for the benzene alkylation reaction.

Secondly, we studied the alkylation reaction of benzene with benzyl chloride at Trea= 80°C in the presence of the following calcined catalysts : CoMgLa, NiMgLa (mcat = 0.1g) with a molar ratio Bz/ClBz = 15.0. The results of this study are shown in Figure-11 and Table-4.

Table-4: The catalytic properties of calcined solids of the MeMgLa type (Me = Co, Ni). The ratio of benzene to benzyl chloride = 15.0. Reaction temperature = 80°C, reaction time = 3h).

CatalystsBenzyl	chloride conversion (%)	Selectivity (%)[a]	
		Diphenylpoly benzyl methane	benzene
MgLa	-	-	-
CoMgLa	19,5	100	-
NiMgLa	22,7	100	-

[a] Selectivity recorded at 3 hrs reaction time.

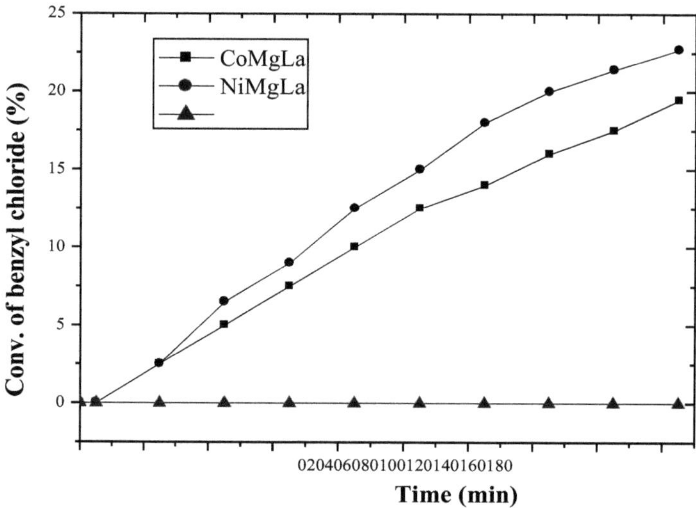

Figure-11: Evolution of the conversion of benzyl chloride as a function of the reaction time at Trea = 80°C of the Me MgLa series with Me = Co, Ni.

We note essentially that the catalysts obtained after the introduction of a transition element (Ni, Co), present a catalytic activity of notable order contrary to the basic support MgLa, which is completely inactive.

In addition, the NiMgLa catalyst shows slightly higher catalytic activity than its CoMgLa counterpart. Nevertheless, a complete selectivity in diphenyl methane (100%) is obtained in the case of these two catalysts.

It should also be noted that MgAl-HTc-based samples have better catalytic performance than MgLa-based samples.

In addition, we compared the best performing hydrotalcite-type FeMgAl-HTc sample to other iron-doped HMS-type mesoporous silicate systems already tested in the benzylation reaction of benzene with benzyl chloride under the same operating conditions [30].

We then noticed that both types of systems have total selectivity for diphenyl methane. On the other hand, the conversion of benzyl chloride is higher than that of the reaction in the presence of our catalyst. Indeed, it goes from 58.3% in our case, to a total conversion of benzyl chloride (100%) for Fe- HMS-50.

We believe that the fact that FeMgAl-HTc is less active than the Fe-HMS- 50 catalyst could be due either to the acidic nature of the HMS system framework or to the low calcination temperature of our samples.

Indeed, these results open up interesting prospects for the continuation of our work.

In order to elucidate the mechanics of this reaction in the presence of our solids, kinetic measurements are undertaken in another Magister dissertation, taking benzene as a model molecule. The parameters that we then set out to study are: the order of the reaction, the reaction molar ratio, the effect of the reaction temperature as well as the effect of the addition of water and the influence of substituents on yield and selectivity, for different aromatic compounds.

BIOGRAPHICAL REFERENCES

[1]-C. Crisafulli, S. Scirè, S. Miniò, L. Solarino, Appl Catal. A: Gen 255 (1991) 1.

[2]-T. Shishido, M. Sukenobu, H. Morioka, R. Furukawa, H. Shirahase, K. Takehira, Catal Lett 73 (2001) 21.

D. Halliche, A. Auroux, O. Cherifi. J.Soc. Alger Chim, 142 (2004) 281.

[4]-A. Tsyganok, T. Tsunoda, S. Hamakawa, K. Suzuki, K. Takehira, T. Hayakawa, J Catal 213 (2003) 191.

5]-P. Gronchi, P. Centola, R. Del Rosso, Appl Catal 152 (1997) 83.

[6]-A. Erdohely, J. Cserenyi, E. Papp and F. Solymosi, Appl Catal. A108 (1994) 205-219. 7]-J. Guo, H. Lou, H. Zhao, D. Chai, X. Zheng, Appl Catal A: Gen 273 (2004) 75.

[8]-N. M. Rodiguez, J. Mater. Res, 8 (1993) 3223.

9-J.T. Richardson, S. A. Paripatyadar, Catalan Application, 61 (1990) 293.

10]-O. W. P. Lopez, A. Senger, N. R. Marcilio, M. A. Lansarin, Appl Catal A: Gen 303 (2006) 234.

[11]-Z. Hou, T. Yashima, Appl Catal A: Gen 261 (2004) 205.

[12]-A. Bhattacharyya, V. W. Chang, D. J. Schumacher, Appl Clay Sci 13 (1998) 317.

[13] Z.Abdelssadek, F.Touhra, K.Bachari, A.Saadi, O.Cherifi, D.Halliche, Adv micro mesop mat, accepted, (2007) in press.

14]-R.E. Reitmeier, K. Atwood , H.A. Bennet, J.r. and H.M. Baugh, Ind. Eng. Chem. 40 (1948) 620.

15]-G.A. White, T.R. Roszkowski, D.W. Stanbridge, Hydrocarbon Process, 54 (1975) 130.

[16]-A. Sacco, Jr, F.W.A.H. Geurts, G.A. Jablonski, S.Lee , R.A. Gatell, J. Catal, 119 (1989) 322.

M. Rodiguez, J. Mater. Res, 8 (1993) 3233.

[18]-C. Bartholomew, Catal. Rev.-Sci. Eng. 24 (1) (1982) 67.

[19]-T. Koerts, R.A. Van Santen, J. Chem. Soc., Chem. Commun, (1991) 1281.

[20]-G. A. Ja blonski, F.W.A.H. Geurts, A. Sacco, R.R. Birdmann, Carbon 30 (1992) 87.

[21]-Z.L. Zhang, X.E. Verykios, Catal. Today,21 (1994) 589.

22]-J. Guo, H. Lou, X. Zheng, Carbon 45 (2007) 1314.

23]-P.Gronchi, P. Centola, P. Centola, R. Del Rosso, Appl Catal A: Gen 152 (1997) 83.

[24]-Z. Hou, O. Yokota, T. Tanaka, T. Yashima, Apll Surface Scien 233 (2004) 58.

[25]-G. A. Olah, Friedel-Craft Chemistry, Wiley, New York, 1973.

[26]-B. Cop, V. Gourves, F. Figueras, Catalan application, 100 (1993) 69.

F. Figuears, T.Cseri, S. Bekassy, S. Rizner, J.Mol. Catal. A: Chem.98 (1995) 101. 28]-
V.R. Choudhary, S.K. Jana, J. Mol. Catal.A, 180 (2002) 267.

29]-V.R. Choudhary, S.K. Jana, B.P. Kiran, Catal.Lett. 64 (2002) 233.

[30]-K. Bachari, J.M.M. Millet, B. Benaïchouba, O. Cherifi, F.Figueras, J Catal 221 (2004) 55.

Conclusion
General

OVERALL FINDING

In this thesis, we are interested in the application of mesoporous hydrotalcite-type materials in the dry methane reforming reaction as well :

- Two series of hydrotalcite-type catalysts, noted M(II)M'(II)M(III)-HT where M(II) = Ni, Co, Fe, Cr, Cu, and M' (II) = Mg and M(III) = Al, La , are prepared by the co-precipitation method at constant pH (pH= 11.0).

- These catalysts have been the subject of various investigations, including chemical analysis, X-ray diffraction, scanning electron microscopy, infrared spectroscopy (FTIR), specific surface area measurement using the BET method and the fluorescence method.

- Chemical analysis of the calcined samples confirmed that the molar ratios M(II) / M(III) = 2, in all cases .

- Examination of the calcined and uncalcined catalysts by X-ray diffraction revealed the formation of the hydrotalcite framework, which disappears in part after calcination of the samples at 450°C.

- Analysis of the uncalcined samples by infrared spectroscopy confirmed the formation of the double lamellar structure of the hydrotalcites. Residual carbonates were detected on the infrared spectra of the calcined samples, allowing the initial structure of the hydrotalcite to be recovered (memory effect).

- All samples were tested in a dry methane reforming reaction. This reaction was studied in a dynamic regime at atmospheric pressure.

- It has been observed that our catalysts reached steady state after 1h 30 min of working under reaction mixing.

- The study of the catalytic performance of the samples, as a function of temperature, showed that, overall, MgAl-HTc hydrotalcite-based catalysts had better catalytic performance compared to MgLa hydrotalcite-based catalysts.

- Another remarkable result concerns cobalt-based catalysts which, under conventional reaction conditions, showed no catalytic activity.

- The study of the effect of in-situ reduction temperature of catalysts has made it possible to re-examine the optimal operating conditions for cobalt-based catalysts.

- Thus the sample is tested under reaction mixture under different reduction experimental conditions, i.e. reduced overnight at 500°C and 650°C. The reactivity of the samples was also examined when the sample was not subjected to any reduction beforehand.

- It has been shown that NiAl-HTc and NiMgAl-HTc solids show good catalytic performance even when the sample is not reduced, in contrast to CoMgAl-HTc catalyst, which does not show any catalytic activity if not reduced at a higher temperature, i.e. 650°C.

- The study of catalyst ageing showed that the deactivation of our samples could be more likely due to the sintering of the active phase.

- The NiAl-HTc catalyst has shown remarkable resistance to deactivation, so further studies are needed to elucidate the mode of operation of this type of catalyst.

- The catalysts based on Cr, Fe and Cu that were totally inactive in the dry methane reforming reaction were then oriented towards another catalytic application of the Friedel-Craft type (alkylation of benzene by benzyl chloride).

- For the latter reaction (Friedel-Craft), catalysts based on MgAl-HTc hydrotalcite have been shown to show very promising results in terms of activity and selectivity.

Table of Contents

Printed by Books on Demand GmbH, Norderstedt / Germany